Springer Series in Materials Science

Volume 299

The Springer Series in Materials Science covers the complete spectrum of materials research and technology, including fundamental principles, physical properties, materials theory and design. Recognizing the increasing importance of materials science in future device technologies, the book titles in this series reflect the state-of-the-art in understanding and controlling the structure and properties of all important classes of materials.

More information about this series at http://www.springer.com/series/856

Bharat Bhushan

Bioinspired Water Harvesting, Purification, and Oil-Water Separation

 Springer

Bharat Bhushan
Department of Mechanical Engineering
The Ohio State University
Columbus, OH, USA

ISSN 0933-033X   ISSN 2196-2812   (electronic)
Springer Series in Materials Science
ISBN 978-3-030-42134-2   ISBN 978-3-030-42132-8   (eBook)
https://doi.org/10.1007/978-3-030-42132-8

This Springer imprint is published by the registered company Springer Nature Switzerland AG
The registered company address is: Gewerbestrasse 11, 6330 Cham, Switzerland

*To my grandkids: Sahana, Ashwin,
Joya and Nivaan*

Preface

Fresh water sustains human life and is vital for human health. It is estimated that about 800 million people worldwide lack basic access to drinking water. About 2.2 billion people (nearly a third of global population) do not have access to a safe water supply, free of contamination. Also, over 2 billion people live in countries experiencing high water stress. Water consumption continues to grow worldwide, driven by a combination of population growth, socioeconomic development, and rising demand in the industrial and domestic sectors. The current supply of fresh water needs to be supplemented to meet future needs. Living nature has evolved species, which can survive in the most arid regions of the world by water collection from fog and condensation in the night. Before the collected water evaporates, species have mechanisms to transport water for storage or consumption. These species possess unique chemistry and structures on or within the body for collection and transport of water.

Water contamination by human activity and unsafe industrial practices, as well as population, continues to grow. Water contamination is one of the major environmental and natural resource concerns in the twenty-first century, including North America. Oil contamination can occur during operation of machinery, oil exploration and transportation, and due to operating environment. Oil spills occasionally occur during oil exploration and transportation. Commonly used oil-water separation techniques in oil spill cleanups are either time consuming, energy intensive, and/or environmentally unfriendly. For water purification and oil-water separation for various applications including oil spill cleanup, bioinspired superhydrophobic/superoleophobic and superoleophobic/superhydrophilic surfaces have been developed which are sustainable and green or environmentally friendly. Bioinspired oil-water separation techniques can be used to remove oil contaminants from both immiscible oil-water mixtures and oil-water emulsions. Coated porous surfaces with an affinity to water and repellency to oil and vice versa are commonly used. Oil-water emulsions require porous materials with a fine pore size.

In this book, an overview of arid desert conditions, desert plants and animals, water harvesting lessons from nature, and water collection data from various designs of bioinspired surfaces from fog and condensation from ambient, are presented.

Next, consumer to emergency and military applications are discussed and various designs for water harvesting towers and projections for water collection are presented. Finally, bioinspired water desalination, water purification, and oil-water separation techniques are described.

The book should serve as an excellent text for a one-semester course in biomimetics, water supply and management, and environmental engineering. The book is also intended for use by novices as well as experts in the field, practitioners, solution seekers, and the curious. Applications should help in advancement of the field.

Contributions by many graduate students, postdoctoral fellows, and visiting scholars in the author's laboratory as well as international collaborators are gratefully acknowledged. The book is largely based on major contributions by the author's postdoctoral fellows—Dr. Philip S. Brown and Dr. Dong Song, graduate students—Dev Gurera, Charles T. Schriner, Wei Feng, and Feiran Li, and an international scholar—Prof. Aditya Kumar. The author would like to thank Dev Gurera for preparing some of the figures and Joanne F. Holland for typing the manuscript. The author would also like to thank his wife Sudha for her constant support and encouragement.

Columbus, USA   Bharat Bhushan
                                                            Bhushan.2@osu.edu

Contents

About the Author

Dr. Bharat Bhushan received an M.S. in mechanical engineering from the Massachusetts Institute of Technology in 1971, an M.S. in mechanics and a Ph.D. in mechanical engineering from the University of Colorado at Boulder in 1973 and 1976, respectively, an MBA from Rensselaer Polytechnic Institute at Troy, NY, in 1980, Doctor Technicae from the University of Trondheim at Trondheim, Norway, in 1990, a Doctor of Technical Sciences from the Warsaw University of Technology at Warsaw, Poland, in 1996, Honorary Doctor of Science from the National Academy of Sciences at Gomel, Belarus, in 2000, University of Kragujevac, Serbia, in 2011, and Honorary Doctorate from the University of Tyumen, Russia. He is Registered Professional Engineer. He is presently an Ohio Eminent Scholar and the Howard D. Winbigler Professor in the College of Engineering, Director of the Nanoprobe Laboratory for Bio- & Nanotechnology and Biomimetics (NLB^2), and Affiliated Faculty in The John Glenn College of Public Affairs at the Ohio State University, Columbus, Ohio. In 2013–2014, he served as ASME/AAAS Science & Technology Policy Fellow, House Committee on Science, Space & Technology, United States Congress, Washington, DC. He has served as Expert Investigator on IP-related issues in the US and international courts. His research interests include fundamental studies with a focus on scanning probe techniques in the interdisciplinary areas of bio-/nanotribology, bio-/nanomechanics, and bio-/nanomaterials characterization and applications to bio-/nanotechnology and biomimetics. He is an internationally recognized expert of bio-/nanotribology and bio-/nanomechanics using scanning probe microscopy and biomimetics. He is considered by someone of the pioneer of the tribology and mechanics of magnetic storage devices, nanotribology, and biomimetics. He is one of the most prolific authors. He has authored 10 scientific books, 90+ handbook chapters, 900+ scientific papers (One of Google Scholar's 1494 Highly Cited Researchers in All Fields (h > 100), h-index—125+ with 75k+ citations, i10-index—740+; Fourth Highly Cited Researcher in Mechanical Eng.; Web of Science h-index—95+; Scopus h-index—100+; ISI Highly Cited Researcher in Materials Science since 2007 and in Biology and Biochemistry, 2013; ISI Top 5% Cited Authors for Journals in Chemistry, 2011; Clarivate Analytics Highly Cited Researcher in Cross-field Category, 2018), and 60+ technical reports. His research was listed as Top Ten Science Stories of 2015. He has also edited 50+ books and holds more than 25 US and foreign patents. He is Co-editor of Springer NanoScience and Technology Series and Co-editor of Microsystem Technologies. He has given more than 400 invited presentations on six continents and more than 200 keynote/plenary addresses at major international conferences. He delivered a *TEDx 2019 lecture* on Lessons from Nature.

His biography has been listed in over two dozen Who's Who books including Who's Who in the World. He has received more than two dozen awards for his contributions to science and technology from professional societies, industry, and US government agencies including Life Achievement Tribology Award and Institution of Chemical Engineers (UK) Global Award. He received NASA's Certificate of Appreciation to recognize the critical tasks performed in support of President Reagan's Commission investigating the Space Shuttle Challenger Accident. He is also the recipient of various international fellowships including the Alexander von Humboldt Research Prize for Senior Scientists, Max Planck Foundation Research Award for Outstanding Foreign Scientists, and Fulbright Senior Scholar Award. He is Foreign Member of the International Academy of Engineering (Russia), Byelorussian Academy of Engineering and Technology, and the Academy of Triboengineering of Ukraine,

Honorary Member of the Society of Tribologists of Belarus and STLE, Fellow of ASME, IEEE, and the New York Academy of Sciences, and Member of ASEE, Sigma Xi, and Tau Beta Pi.

He is an accomplished organizer. He organized the Ist Symposium on Tribology and Mechanics of Magnetic Storage Systems in 1984 and the Ist International Symposium on Advances in Information Storage Systems in 1990, both of which are now held annually. He organized two international NATO institutes in Europe. He is Founder of an ASME Information Storage and Processing Systems Division founded in 1992 and served as Founding Chair during 1993–1998.

He has previously worked for Mechanical Technology Inc., Latham, NY; SKF Industries Inc., King of Prussia, PA; IBM Tucson, AZ; and IBM Almaden Research Center, San Jose, CA. He has held visiting professorship at University of California at Berkeley; University of Cambridge, UK; Vienna University of Technology, Austria; University of Paris, Orsay; ETH Zurich; EPFL Lausanne; University of Southampton, UK; University of Kragujevac, Serbia; Tsinghua University, China; Harbin Institute of Technology, China; and KFUPM, Saudi Arabia.

https://nlbb.engineering.osu.edu/
https://www.facebook.com/bhushanb100

Chapter 1
Introduction: Water Supply and Management

The Human Right to Water: UN General Assembly Resolution 64/292 (2010) "recognizes the right to safe and clean drinking water and sanitation as a human right that is essential for the full enjoyment of life and all human rights."

11th Annual Cisco Visual Networking Index Global Mobile Data Traffic Forecast. By 2021, more members of the global population will be using mobile phones (5.5 billion) than running water (5.3 billion).

1.1 Water Supply

Access to a safe water supply is a human right. Fresh water sustains human life and is vital for human health. Water availability depends upon the amount of water physically available, and whether it is safe for human consumption. Some of the arid regions of the world lack adequate safe drinking water. Figure 1.1 shows proportion of population with access to safe drinking water in 2015 (WHO/UNICEF 2017; WWAP 2019; Bhushan 2020). A total of 181 countries had a population of over 75% with different degrees of accessibility (Bhushan 2020). Due to bad economies or poor infrastructure in some parts of the world, water accessibility continues to become even worse particularly in these regions.

It is estimated that about 800 million people worldwide lack basic access to drinking water (WHO/UNICEF 2017; Bhushan 2020). This means that they cannot reach a protected source of drinking water within a total walking distance of 30 min. About 2.2 billion people, representing nearly one third of the world population do not have access to a safe water supply, meaning no drinking water on the property that is available at all times and free of contamination. A lack of drinking water particularly affects rural areas; 80% of the people lacking drinking water live in rural areas. Almost half of people drinking water from unprotected sources live in sub-Saharan Africa. Women and girls regularly experience discrimination and inequalities in the rights to safe drinking water. Ethnic and other minorities, disability, age and health status are also factors.

© Springer Nature Switzerland AG 2020
B. Bhushan, *Bioinspired Water Harvesting, Purification, and Oil-Water Separation*, Springer Series in Materials Science 299, https://doi.org/10.1007/978-3-030-42132-8_1

Proportion of population with access to drinking water services in 2015

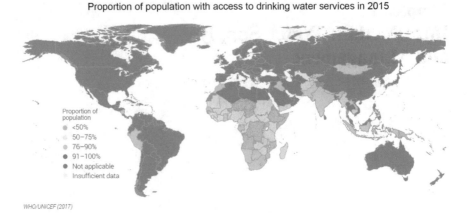

WHO/UNICEF (2017)

Fig. 1.1 Proportion of population with access to basic drinking water services in 2015. 181 countries had a coverage of over 75% (adapted from WHO/UNICEF 2017; Bhushan 2020)

Level of physical water stress

*Physical water stress is defined here as the ratio of total freshwater withdrawn annually by all major sectors, including environmental water requirements, to the total amount of renewable freshwater resources, expressed as a percentage (UN, 2018)

Fig. 1.2 Level of physical water stress showing percent of total fresh water withdrawn annually (adapted from UN 2018; Bhushan 2020)

Because of global warming and an increased number of drought periods, water supply continues to shrink. Figure 1.2 shows the level of physical water stress around the world showing percent of total fresh water withdrawn annually (UN 2018; Bhushan 2020). Over 2 billion people, representing nearly one third of the world population, live in countries experiencing high water stress, meaning that water resources consumed are not regenerated to the necessary extent by rain or the return of the purified water (UN 2018). About 4 billion people experience severe water scarcity during at least one month of the year (WWAP 2019). Stress levels will continue to increase as demand of water grows and the effects of climate change intensify. The majority of the people affected by water stress live in North Africa, the Middle East, and South Asia, but the people living in North and South America including the southwest United States are also increasingly being affected.

Figure 1.3a shows a child drinking contaminated water from the river (adapted from image provided by Water Crisis in Africa). In many parts of the world, people need to transport water from far which is often not so clean (Fig. 1.3b) (adapted from Left-image provided by Picsio, Creative Commons, and Right-image provided by Reuters). It is often children and women who do most of the water transporting (Bhushan 2020).

Roughly, 70% of the Earth's surface is covered by water, however, the vast majority of water is contained in the oceans with fresh water accounting for only about 2.5% of all water, as shown in Fig. 1.4a (Brown and Bhushan 2016; Bhushan 2018, 2019, 2020). Water continuously moves in a cycle due to evaporation, condensation, precipitation, surface and channel runoff, and subsurface flow, as shown in Fig. 1.4b (Brown and Bhushan 2016). The evaporative phase can help purify water by separating it from contaminants picked up in other phases of the cycle, including salt in the oceans. Of the 2.5% fresh water, the majority is trapped as ice in glaciers and snow

Lack of clean water

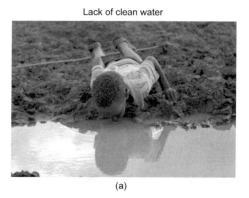

(a)

A long walk for access to clean water

(b)

Fig. 1.3 **a** A child drinking contaminated water from the river, **b** people transporting water from far (adapted from Bhushan 2020)

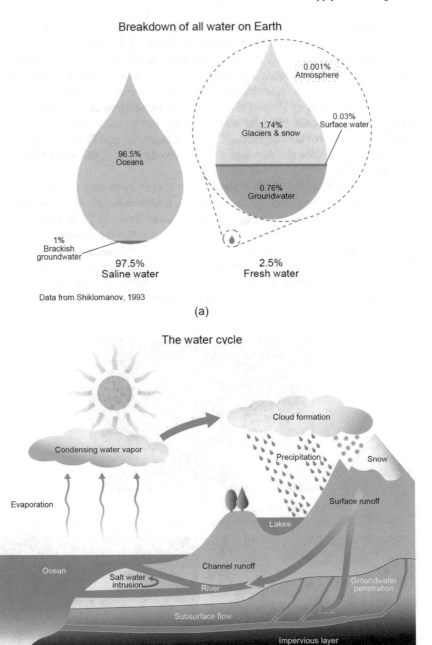

Fig. 1.4 **a** Percentage breakdown of all water on Earth. 2.5% is fresh water, with surface water in rivers and lakes only accounting for 0.03% of all water (400 trillion m^3) and groundwater only 0.79% of all water (11 quadrillion m^3) (Shiklomanov 1993), and **b** the water cycle where of all water the evaporative phase helps purify water via separation from contaminants picked up in other phases of the cycle (adapted from Brown and Bhushan 2016; Bhushan 2018, 2019, 2020)

(1.74% in total). Only 0.79% (11 quadrillion m^3) of all water is found as surface water in lakes and rivers (0.03% or 400 trillion m^3), and as groundwater (0.76%) (Shiklomanov 1993). The distribution of this water is not uniform across the world, with around 20% of the world's surface fresh water found in the North American Great Lakes (EPA 1995).

Given that about 97.5% of water is saline water, desalination has become increasingly important in all parts of the world. Ocean water is contaminated by salt, as well as by bacteria and particulates. For human consumption, ocean water must be desalinated and purified. However, desalination remains an energy intensive process and is prohibitively expensive (Elimelech and Phillip 2011). Desalination may consume as much as 30 times more energy than other water purification approaches. Furthermore, for every 50 L of seawater, desalination produces about 1 L of clean water as well as 2 L of brine solution which poses serious environmental issues. Globally, desalination provides a small fraction of purified clean water supply. In 2020, one fourth of all global desalination of water was carried out in Saudi Arabia and U.A.E. In the United States, Carlsbad Desalination Plant in California is the largest desalination plant. It delivers about 200,000 L of fresh desalinated water to San Diego County, enough to serve about 400,000 people.

In addition to all water on earth, the earth's atmosphere contains about 0.001% of global water (on the order of 13 trillion m^3 or 1.3×10^{16} L). It is found in the form of clouds, fog, mist, and water vapor. It is a small amount in global terms but it is about 3.3% of water in rivers and lakes which can be a major source of drinking water. In arid deserts, many plants and animals survive by harvesting water from the atmosphere. The water is a recyclable natural resource with potential to water the arid regions of the world (Gleick 1993; Bhushan 2018). Scientists and engineers have been working to develop bioinspired water harvesters.

1.2 Water Consumption

About 60% of body weight in adults and 75% in infants is water. The body constantly loses water. It needs to be replenished by drinking water. Water is needed for food preparation and sanitation and personal hygiene. A summary of water consumption by humans is presented in Table 1.1 (Day et al. 2019; Bhushan 2020). The amount of water required to meet basic human water needs is about 7.5–15 L day^{-1}, with 7.5 L day^{-1} being the absolute minimum of the individual's water needs (Day et al. 2019; Bhushan 2020). In the U.S.A., about 600 L of water per capita and per day is consumed; whereas, in Europe it is about 300 L of water per capita and in the rest of the world is much less. Typical water requirements per capita is on the order of 50 L day^{-1}. It is noted that these numbers do not include water requirements for food production and other consumer needs, which can be much higher than water consumption by humans.

Water consumption continues to grow worldwide, driven by a combination of population growth, socio-economic development, and rising demand in the

Table 1.1 Water consumption by various processes by humans (adapted from Day et al. 2019; Bhushan 2020)

Process	Amount of water lost/used (L day^{-1})
Body constantly releases water through evaporation from the skin, through urine production and through the release of water from the lungs with the air we breathe. It needs to be replenished by drinking water	~2.4
Food preparation	3–6
Sanitation and personal hygiene practices	2–6
Total basic needs	**7.5–15**
Typical needs worldwide	**50**

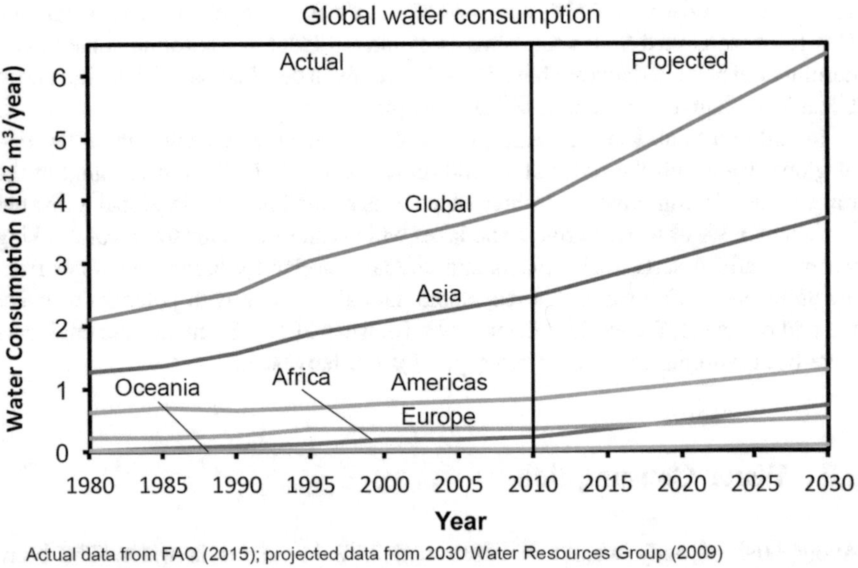

Actual data from FAO (2015); projected data from 2030 Water Resources Group (2009)

Fig. 1.5 Global water consumption based on actual (1980–2010) data (UN 2015) and projected (2030) data (adapted from Water Resources Group 2009; Brown and Bhushan 2016; Bhushan 2018, 2019)

industrial and domestic sectors. With ever-growing population, there is an increasing demand for food and other necessities. Food production requires a significant amount of water. Water consumption has been increasing by about 1–2% per year since the 1980s (WWAP 2019). Global water consumption is expected to reach 6 trillion m^3 a year by 2030 as the population moves toward a projected 8.2 billion, as shown in Fig. 1.5 (Water Resources Group 2009; UN 2015; Brown and Bhushan 2016; Bhushan 2018, 2019).

1.3   Water Contamination

Water contamination by human activity and unsafe industrial practices, as well as population growth continues to grow. Water contamination is one of the major environmental and natural resource concerns in the twenty-first century (Clark et al. 1986; Rao et al. 2012; Brown and Bhushan 2016; Fingas 2017; Bhushan 2018). Contamination by oil can occur during operation of machinery, oil exploration and oil transportation, and due to operating environment. Oil is commonly used as a lubricant in machinery and can leak and produce contamination. Oil spills occasionally occur during oil exploration and transportation. They will continue to occur as long as society is dependent upon oil (Brown and Bhushan 2016; Bhushan 2018). The Deepwater Horizon oil spill in the Gulf of Mexico in 2010 was the worst accidental offshore oil spill in history. It spilled about 200 million U.S. gallons of oil (Fig. 1.6). The partial cleanup by British Petroleum took several years and full cleanup was never accomplished. Oil spills cause damage to the environment and contaminate sources of potable water. Another source of contamination stems from the emergence of fracking in the U.S., where water-based fluids (containing sand and chemicals) are injected under high pressure to fracture rocks to release oil and gas, which lead to oil-contaminated wastewater.

Water contamination with a variety of chemicals is another major concern with growing world population and industrialization with unsafe practices of waste disposal. Various organic contaminants present in contaminated water pose health risks.

Access to "clean" and safe water is a global problem. While the situation is particularly severe for large number of countries in Africa, Asia, and the Middle East, Latin America, Caribbean, and some countries in Eastern Europe, many people in Western and Coastal Europe as well as in North America also suffer from lack of inequitable access to safe drinking water (WWAP 2019).

Deepwater Horizon oil spill, Gulf of Mexico (2010)

Fig. 1.6 Photograph of oil-contaminated water after the 2010 Deepwater Horizon oil spill in the Gulf of Mexico

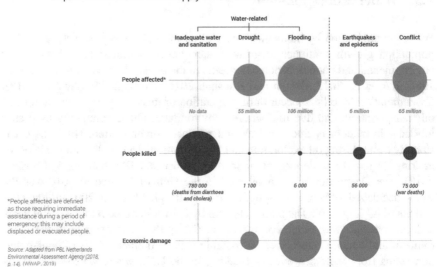

Fig. 1.7 Average annual impact from inadequate drinking water and sanitation services, water related disasters, epidemics and earth quakes and conflicts (adapted from WWAP 2019)

1.4   Human Impact from Lack of Safe Water Supply

Water contamination is a major health issue facing the world today. Lack of safe water supply takes a toll on living nature—flora, fauna, and humans. It leads to severe diseases and even death, as well as economic damage. Contaminated drinking water can transmit diseases such as diarrhea, cholera, dysentery, typhoid, and polio. It is estimated to cause some 500,000 diarrheal deaths each year. Figure 1.7 shows the impact of lack of safe water supply as well as natural disasters and conflicts (WWAP 2019). The impact of inadequate water and sanitation and drought affect many people with significant economic damage. Therefore, water purification becomes increasingly important.

1.5   Lessons from Nature to Supplement Water Supply and Remove Contamination

It is apparent that the current supply of fresh water needs to be supplemented to meet future needs. To find new sources of water supply, living nature may provide solutions. In living nature, after some 3.8 billion years of evolution, many plant and animal species in arid regions exhibit efficient solutions for water harvesting from fog and condensation of water vapor. These solutions typically involve species possessing unique surface structures and chemistry on or within their bodies that help to direct the movement of water to where it is either consumed or stored, before it is evaporated (Brown and Bhushan 2016; Bhushan 2018, 2019, 2020).

By studying the surface structures and chemistries involved, bioinspired fog harvesters are being developed (Bhushan 2019, 2020). The ambient temperature during desert nights is low, as low as 0–4 °C. This temperature can be lower than the dew point, which would lead to water condensation of water vapor from ambient. Bioinspired water condensation can also be used for water harvesting (Bhushan 2018, 2019, 2020).

Bioinspired water harvesters can be used to provide a supplemental water source for communities in the arid regions where fog and condensation of water vapor is common, such as the coastal regions of Africa, the Southwestern coast of South America, and the Southwestern United States (Brown and Bhushan 2016; Bhushan 2018, 2019, 2020). In addition, these harvesters can be used in various emergency and defense applications (Bhushan 2019, 2020). Emergency applications, such as natural disasters, could benefit for short periods from portable units which could be dropped from air. The cost of clean water in military bases in a combat zone can be high, and water harvesters become attractive. Various designs for large and portable 3D towers have been developed.

Systems in human bodies have evolved to transport water efficiently while blocking other molecules and ions. Inspiration can be taken to improve the efficiency of desalination and help purify contaminated water. Using inspirations from Lotus leaf and other species, a coated mesh which repels oil and attracts water or vice versa, can be used for oil-water separation. Various approaches for bioinspired water purification and oil-water separation have been developed (Brown and Bhushan 2016; Bhushan 2018).

Bioinspired approaches are attractive to develop materials and surfaces in an environmentally friendly and sustainable manner to supplement water supply and remove contamination.

1.6 Organization of the Book

In this book, an overview of bioinspired water harvesting methods from fog and water condensation of water vapor from ambient, to supplement water supply and various water purification and oil-water separation approaches, are presented. First, an overview of arid desert conditions, water sources, and desert plants and animals is presented. Next, water harvesting mechanisms of selected plants and animals, which are being exploited for commercial applications are presented with a discussion of the surface structures and chemistries used. Then, water collection data for various designs of bioinspired surfaces, both from fog and condensation from ambient, are presented, which can be used to develop optimal designs. Various commercial, emergency and military applications are discussed and various designs for water harvesting towers and projections for water collection are presented.

Finally, bioinspired water desalination, water purification and oil water separation techniques are described. All bioinspired approaches described in the book can be used to develop materials and surfaces in an environmentally friendly and sustainable manner to supplement water supply and remove contamination.

References

Bhushan, B. (2018), *Biomimetics: Bioinspired Hierarchical-Structured Surfaces for Green Science and Technology*, third ed., Springer International, Cham, Switzerland.

Bhushan, B. (2019), "Bioinspired Water Collection Methods to Supplement Water Supply," *Phil. Trans. R. Soc. A* **377**, 20190119.

Bhushan, B. (2020), "Design of Water Harvesting Towers and Projections for Water Collection from Fog and Condensation," *Phil. Trans. R. Soc. A* **378**, 20190440.

Brown, P. S. and Bhushan, B. (2016), "Bioinspired Materials for Water Supply and Management: Water Collection, Water Purification and Separation of Water from Oil," *Phil. Trans. R. Soc. A* **374**, 20160135.

Clark, R. B., Frid, C., and Attrill, M. (1986), *Marine Pollution*, Vol. 4, Oxford University Press, Oxford, U. K.

Day, S. J., Forster, T., Himmelsbach, J., Korte, L., Mucke, P., Radtke, K., Theilbörger, P., and Weller, D. (2019), *World Risk Report 2019 – Focus: Water Supply*, Bündnis Entwicklung Hilft, Berlin, Germany.

Elimelech, M. and Phillip, W. A. (2011), "The Future of Seawater Desalination: Energy, Technology, and the Environment," *Science* **333**, 712–717.

EPA (1995), *The Great Lakes: An Environmental Atlas and Resource Book 3rd Edition*, Environmental Protection Agency, Chicago, Illinois.

Fingas, M. (2017), *Oil Spill Science and Technology*, second ed., Gulf Professional Publishing, Houston, Texas.

Gleick, P. H. (1993), *A Guide to the World's Fresh Water Resources*, Oxford University Press, Oxford, U. K.

Rao, D. G., Senthilkumar, R., Byrne, J. A., and Feroz, S. (2012), *Wastewater Treatment: Advanced Processes and Technologies*, CRC Press, Boca Raton, Florida.

Shiklomanov, I. A. (1993), "World Fresh Water Resources." In *Water in Crisis: A Guide to the World's Fresh Water Resources* (P. H. Gleick, ed.), pp. 13–24, Oxford University Press, Oxford, U. K.

UN (2015), "Water Uses," Food and Agricultural Organization of the United Nations, see http://www.fao.org/nr/water/aquastat/water_use.

UN (2018), *Sustainable Development Goal 6: Synthesis Report 2018 on Water Sanitation*, United Nations, New York. www.unwater.org/app/uploads/2018/07/SDG6_SR2018_web_v5.pdf.

Water Resources Group (2009) "Charting Our Water Future: Economic Frameworks to Inform Decision-making," see http://www.2030wrg.org/wp-content/uploads/2014/07/Charting-Our-Water-Future-Final.pdf.

WHO/UNICEF (2017), *Progress on Drinking Water, Sanitation and Hygiene: 2017 Update and SDG Baselines*, World Health Organization (WHO) and the United Nations Children's Fund (UNICEF), Geneva, Switzerland. See https://washdata.org/sites/default/files/documents/reports/2018-01/JMP-2017-report-final.pdf.

WWAP (2019), *The United Nations World Water Development Report 2019: Leaving No One Behind*, WWAP (UNESCO World Water Assessment Programme), UNESCO, Paris, France.

Chapter 2
Overview of Arid Desert Conditions, Water Sources, and Desert Plants and Animals

Deserts have fascinated humans for centuries. They have been focus of even fictional books (Herbert 1965). A desert is a barren landscape where little precipitation occurs (Meigs 1953; Walker 1992; Costa 1995; Mares 1999; Harris 2003; Allaby 2006; Laity 2008; Greenberger 2009). A desert is referred to as an area of land that receives no more than 250 mm of precipitation per year. The amount of evaporation in a desert often greatly exceeds the annual rainfall, which results in a moisture deficit over the course of a year. All deserts are arid or dry. There is little water available for any living nature to survive. Consequently, living conditions are hostile for living nature including plant and animal life.

Deserts are formed by weathering processes as large variations in temperature between day and night put strains on the rocks which consequently break in pieces. The desert soil is sandy with stones and rocks with low content of organic material and water storage capacity. Deserts cover about one-third of the earth's land surface area or about 10% of the total earth surface. They are home to around 1 billion people—about one-eighth of the Earth's population.

Figure 2.1 shows the map of the deserts of the world (the 7 continents). The world's deserts can be divided into four types based on location: subtropical, coastal, cold, and polar (Walker 1992; Costa 1995; Mares 1999; Harris 2003; Allaby 2006; Laity 2008; Greenberger 2009). The deserts can also be classified as hot deserts and cold deserts. Table 2.1 presents types and locations of deserts and details on their aridness and climate as well as names of major deserts.

Hot deserts include subtropical deserts (inland) and coastal deserts, with subtropical being the hottest. These are also the most arid. Hot deserts can be covered by sand, rock, salt lakes, stony hills and even mountains. Non-polar deserts are hot in the day and chilly at night. The average temperature in the daytime can reach up to 50–60 °C in the summer, and dip to as low as 0 °C or even lower at night time in the winter. The soil surface temperature can reach up to 70 °C.

Cold deserts include cold deserts and polar deserts (Table 2.1), with polar deserts being colder. Cold deserts may be covered with snow or ice but some are so dry that the ice sublimates away. Some cold deserts have a short season of

© Springer Nature Switzerland AG 2020
B. Bhushan, *Bioinspired Water Harvesting, Purification, and Oil-Water Separation*, Springer Series in Materials Science 299,
https://doi.org/10.1007/978-3-030-42132-8_2

World map of deserts (The 7 continents)

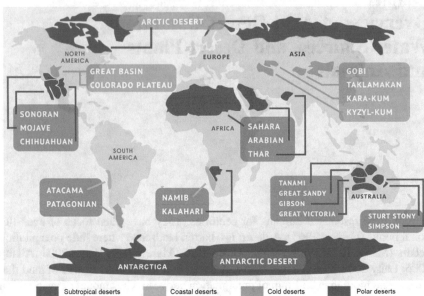

(Adapted from www.whatarethe7continents.com/deserts-of-the-world)

Fig. 2.1 Map of deserts of the world (the 7 continents). They can be divided into four types based on location: subtropical, coastal, cold and polar deserts (adapted from https://www.whatarethe7continents.com/deserts-of-the-world/)

Table 2.1 Types of desert of the world (the 7 continents), based on location and their aridness and climate as well as names of major deserts (adapted from https://www.whatarethe7continents.com/deserts-of-the-world/)

Types and location	Aridness and climate	Names
Hot		
Subtropical deserts—land locked (inland)	Very arid—hot to very hot temperatures with cooler winters	Sahara Arabian Kalahari Mojave Sonoran Chihuahuan Thar Gibson Great Sandy Great Victoria Tanami Sturt Stony Simpson

(continued)

Table 2.1 (continued)

Types and location	Aridness and climate	Names
Coastal deserts—near large bodies of water and/or mountainous regions	Harsh and very arid regions—prolonged moderate summers and mild, cool winters	Namib Atacama
Cold		
Cold deserts—at higher altitudes and rain shadows of high mountains	Semi-arid regions—not as hot as the subtropical deserts and as cold as polar deserts	Great Basin Colorado Plateau Patagonian Karakum KyzlKum Taklamakan Gobi
Polar deserts—near north and south poles	Arid regions—mild with temperatures less than 10 °C	Arctic Antarctic

World map of deserts with non-polar and arid land

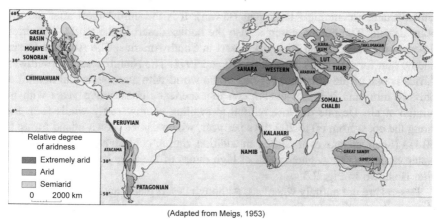

(Adapted from Meigs, 1953)

Fig. 2.2 World map of deserts with non-polar and arid land with relative degree of aridness. Hot and most arid deserts exist in parts of Africa, the Middle East, Southwestern South America and Southwestern United States (adapted from Meigs 1953)

above-freezing temperatures. An ice cap can be a cold desert that remains freezing year round. The annual precipitation is generally less than 50 mm in the interior. The polar deserts which are the coldest, include Antarctica desert located around the South Pole and Arctic desert located above 75° north latitude. Antarctica is the world's largest desert (14 million km^2), closely followed by Arctic (13,985 km^2).

A map of the world's deserts with non-polar arid land, with relative degrees of aridness is shown in Fig. 2.2. The largest (non-polar) hot desert in the world is the Sahara in North Africa. It has one of the harshest environments on Earth, covering some 5.8 million km^2, nearly a third of the African continent, about the size of the

High dunes of the Namib desert near Sossus Viei

Fig. 2.3 High dunes of the Namib desert near Sossus Vlei

United States (including Alaska and Hawaii). It is the third largest after the two polar deserts. The Sahara desert is also the hottest desert in the world. The driest desert in the world is the Atacama Desert in South America. The Atacama Desert had no rain for 401 years, between 1570 and 1971. The Namib desert in southwest Africa is one of the most arid regions in the world with average annual rainfall of only 18 mm, and it is not uncommon to experience consecutive years with no rainfall at all. Despite the low rainfall, prevailing southwesterly winds form fog along the coast from 60 to 200 days per year, which can be blown inland for up to 50 km (Shanyengana et al. 2002). In addition, during cold nights, condensation of water vapor occurs. A photograph of high dunes of the Namib desert near Sossus Vlei is shown in Fig. 2.3.

Deserts are not entirely waterless. Although desert plants and animals may have to go without rainwater for several years at a time, they can harvest water from ambient atmosphere. In living nature, after some 3.8 billion years of evolution, many plant and animal species exist in arid regions, and exhibit efficient solutions for water harvesting (Cloudsley-Thompson and Chadwick 1964; Axelrod 1979; Costa 1995; Anonymous 1996; Mares 1999; Van Rheede van Oudtshoorn and Van Rooyen 1999; Harris 2003; Bredeson 2009; Murphy 2012; Brown and Bhushan 2016; Bhushan 2018, 2019, 2020; Gurera and Bhushan 2020). Water harvesting solutions typically involve species possessing unique surface structures and chemistry on or within their bodies that help to harvest and direct the movement of water before it evaporates, to where it is consumed or stored.

Based on estimates by the World Wildlife Fund, some 500 species of plants, some 70 known mammalian species, some 90 avian species, some 100 reptilian species, many species of spiders, scorpions and other small arthropods are known to exist in deserts. Many plants and animals are found even in extremely arid regions, for example, in the Namib desert. Table 2.2 presents a list of various plants and

Table 2.2 Numbers and proportions of known endemic and near-endemic taxa found in Namib desert (adapted from Simmons et al. 1998; Gurera and Bhushan 2020)

	Endemics	All species	Endemism (%)
Plants	683	4334	17
Mammals	14	200	7
Birds	14	644	2
Reptiles	59	250	24
Fish	3	113	3
Frogs	6	51	12
Insects	1541	6331	24
Arachnids (spiders, scorpions, etc.)	164	1331	12

Fig. 2.4 Selected examples of desert plants and animals (adapted from Bhushan 2020)

animals with their degree of endemism in the Namib desert (Simmons et al. 1998; Gurera and Bhushan 2020). Various animals include mammals, birds, reptiles, fish, frogs, insects, and arachnids. As an example, photographs of two plants and two animals surviving in deserts are shown in Fig. 2.4 (adapted from, Top left-image provided by Gentry George, Creative Commons, Bottom left-Anderson sady, Creative Commons, and Bottom right-Stu images, Creative Commons) (Bhushan, 2020).

2.1 Water Sources

The species surviving in the deserts need water to survive. Water appears in deserts from various sources including rivers, lakes, oases, groundwater used by roots in the case of plant species, metabolic water in the case of animals, rainwater, snow, fog, mist, and condensation of water vapor. Some desert plants have adapted to harsh environments by growing deep roots that can gather water from several feet under the surface. Some desert animals depend upon metabolic water to various degrees. Fog, mist and condensation as water sources are reliable, whereas rainfall is unpredictable. Fog precipitation can exceed rainfall by about sevenfold at the coast and by two times inland (Shanyengana et al. 2002). In arid and semi-arid regions, fog dominates as a water source.

During the cold nights, particularly in the coastal desert regions, fog, mist and water vapor exist in the atmosphere. Fog and mist consist of water droplets suspended in air which can be harvested and coalesced. In humid climates, air consists of a significant amount of water vapor which can be nucleated into droplets (dew droplets) on cold surfaces and coalesced. It is referred to as condensation of water vapor.

2.1.1 Fog and Mist

Fog and mist occur when the water vapor concentration in the atmosphere reaches saturation (Bhushan 2019). These are a visible aerosol consisting of tiny water droplets suspended in air, which is found close to the land surface in cold nights (Fig. 2.5). They are more frequent near lakes or oceans. Fog cover is formed when warm moist air above the desert land encounters the cooler surface of land or blows over the cold water in the ocean, which changes the invisible gas to tiny visible droplets. The formation is aided by the presence of nucleation sites on which the suspended water phase can coagulate. As an example, small particulates can enhance the fog or mist formation. There is less coalescence of water droplets in

Fig. 2.5 Photographs of fog near coastal desert (adapted from Bhushan 2019)

Fog

mist which makes it less dense and quicker to dissipate. The difference between fog and mist is in their location and density. Mist is referred to as cloud cover, suspended in air, and normally found above mountains. Whereas, fog is a cloud that reaches the ground level, normally found above a water surface. In addition fog is denser and lasts for a longer period and mist is thinner and one can see more clearly through it. Visibility in fog is less than about 1 km; whereas in mist, it is 1–2 km.

Fog consists of small water droplets whose size, typically, are on the order of 10–50 μm, with a concentration on the order of 10–100 microdroplets per cm^3 (Anonymous 1999; Mares 1999). Air in fog has a relative humidity, generally above 95%. Density of water in fog is on the order of 0.05–0.5 g m^{-3}. For reference, density of water is 1000 kg m^{-3}. This source of water is clean from impurities (Basu et al. 2018), which is a great benefit of water collected from fog harvesting.

The number of fog days and amount of fog depends upon the geographical location of the desert as well as the offshore ocean currents. Thick fog is formed near coastal deserts (Meigs 1966; Mares 1999). In coastal regions, such as in the high desert region of the Mohave desert, fog cover can extend inland up to some 75 km, even during the day. Table 2.3 presents data for occurrence period and duration of fog in very arid central Namib desert during a 12 month period (Seely 1979). The fog occurred during 11 months for different number of days and time periods.

During low rainfall months, many species survive on fog. Mean monthly rainfall and hourly fog water collection in the California central coast (Big Sur, CA) is presented in Fig. 2.6, based on two years of fog collection and thirty years of rainwater collection (Hiatt et al. 2012). Fog deposition occurred primarily during the summer months of June through October. The figure also shows that as rainfall decreased in the summer months, fog deposition increased, and as the region receives

Table 2.3 Occurrence period and duration of fogs at Central Namib desert during a 12-month period (adapted from Seely 1979)

	No. of fog days	Duration of fog (h)		Time fogs begin	
		Mean	Range	Earliest	Latest
1977					
September	9	3.8	3–7	24:00	06:00
October	7	3.0	1–5	02:00	06:30
November	11	3.2	1–5	12:30	06:30
December	9	3.4	1–7	01:00	06:00
1978					
January	12	3.6	1–8	22:00	06:00
February	4	3.2	1–5	23:00	08:00
March	5	1.2	1–2	05:00	07:00
April	8	2.1	1–7	02:00	07:00
May	2	2.5	1–4	04:00	06:00
June	0	–	–	–	–
July	4	4.2	1–7	02:00	07:00
August	7	2.9	1–7	24:30	07:00

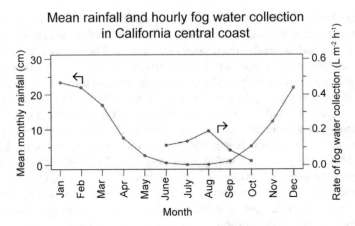

Fig. 2.6 Mean monthly rainfall for years 1981–2010 and hourly fog water collection for the 2010 and 2011 fog collection periods in the California central coast (Big Sur, CA) (adapted from Hiatt et al. 2012)

rainfall in fall, fog deposition decreases. Based on Hiatt et al. (2012), in 2010, a daily average of fog water collection was about 2.3 L m^{-2} and a daily maximum of about 13 L m^{-2}, and an hourly average of about 0.1 L m^{-2} and an hourly maximum of about 2.3 L m^{-2}. Fog deposition occurred during the night and early morning hours. Average fog water collection was about 0.16 L m^{-2} between 8:00 PM and 2:00 AM, increasing to a peak of about 0.2 L m^{-2} between 3:00 AM and 9:00 AM. After 9:00 AM, it steeply declined to nearly 0 between 2:00 PM and 6:00 PM.

Figure 2.7 shows the rate of fog water collected in different regions and the number of days fog occurs annually (Fessehaye et al. 2014). The rate of fog water collected ranges from 1.5 to 12 L m^{-2} day^{-1}. The fog occurrence in the less arid central California coast desert is reported to be more frequent, on the order of 80% of days per year (Hiatt et al. 2012). Fog water collection data in the Central Namib desert was measured with a cylindrical wire mesh at three locations from October 1996 to October 1997. The water collection data is presented in Fig. 2.8 (Shanyengana et al. 2002). The most fog occured from August to February. The daily average collection rate was between 0.5 and 3 L m^{-2} day^{-1} at these locations.

The rate of water collection depends on various environmental factors including the difference between air temperature and dew point temperature, wind speed, wind direction, and land temperature (Hiatt et al. 2012). The difference between air temperature and dew point temperature, termed as the dew point depression, is a strong indicator of fog. Lower is the dew point depression, higher is the fog density, and vice versa. The wind speed of fog is typically on the order of a few cm s^{-1} (United Nations Environment Programme 1999) to as high as a couple of m s^{-1} in the California central coast (Hiatt et al. 2012). Fog water collection is a function of wind speed during prevailing winds. An example of data collected in the California central coast is shown in Fig. 2.9 (Hiatt et al. 2012). Higher is the wind speed, higher is fog collection. Though the wind direction is random, wind should be perpendicular to the collecting surface for maximum collection.

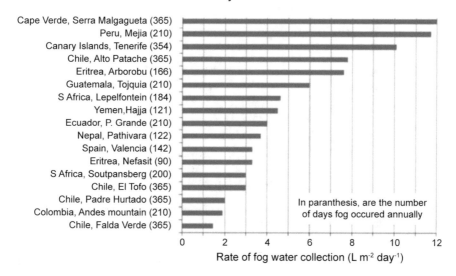

Fig. 2.7 Rate of fog water collection and number of days fog occurs annually at various desert locations (adapted from Fessehaye et al. 2014)

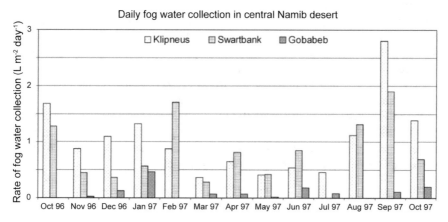

Fig. 2.8 Daily fog collection measured with Standard Fog Collectors in deserts at Gobabeb, Klipneus and Swartbank in central Namib desert, from October 1996 to October 1997 (adapted from Shanyengana et al. 2002)

In summary, average fog water collection is on the order of 2 L m^{-2} day^{-1} or 0.1 L m^{-2} h^{-1}, dependent upon the location. The highest fog collection occurs when the temperature of the land is hot, during the summer months of June and July, and during nights and early mornings. As a reference, a consumer humidifier

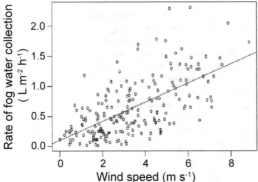

Fig. 2.9 Hourly fog water collection as a function of wind speed during prevailing winds in the California central coast (adapted from Hiatt et al. 2012)

produces fog with water content on the order of 50–1000 L m^{-2} h^{-1} with nozzle diameter on the order of about 20 mm. The fog emitted by the humidifier gets dissipated over a large area.

The fog is one of the major reasons for many species' survival in arid regions. Examples of water harvesting data of one plant and one animal are presented next. Data for the weight gain for tenebrionid beetles (13–22 mm) after a head-down stance to collect water is presented in Table 2.4 (Hamilton and Seely 1976). The beetles, being mobile, consume fog water in two different ways: by consuming fog-water droplets condensed on different vegetation, stones, and soil, and by using a fog-basking stance and drinking the fog-water condensed on its body (Seely 1979). The water collected on *S. sabulicola*, Namib dune bushman grass is presented in Fig. 2.10, measured at 4:00 AM, a few hours after the onset of the fog, and at 8:00 AM. After 8:00 AM, no substantial fog collection could be measured. Measurements were taken on 10 fog events for 5 set of samples (Ebner et al. 2011). Total fog collected per leaf area was on the order of 5 L m^{-2}.

Table 2.4 Weight gain expressed as a percentage weight change of body weight of tenebrionid beetle after a head-down stance to collect water, in fogs of various strengths on the Namib desert (adapted from Hamilton and Seely 1976)

Date	Fog (mm)	Mean weight gain (%)
October 14, 1975	0.50	14.50
November 5, 1975	0.35	1.63
November 25, 1975	0.10	5.47
November 29, 1975	0.65	8.39
December 1, 1975	0.10	2.11

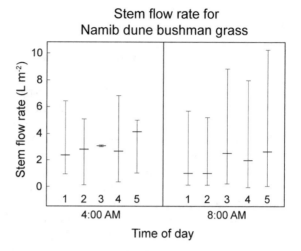

Fig. 2.10 Measured stem flow rates for *S. sabulicola*, Namib dune bushman grass, measured at 4:00 AM and 8:00 AM. Measurements were taken on 10 fog events for five plants. After 8:00 AM, no substantial fog harvesting could be observed. The vertical lines indicate the highest and lowest values and the cross line indicates the median (adapted from Ebner et al. 2011)

2.1.2 Condensation of Water Vapor

Dew occurs as a result of phase transition, in which water vapor in air is transformed into liquid when it comes in contact with a surface. The primary condition for the formation of dew is that the temperature of the surface on which condensation takes place should be lower than the dew point temperature. During phase transition or condensation, nucleation of the liquid phase occurs. Nucleation is the formation of the nanoscale water droplets that are thermodynamically stable (Sigsbee 1969; Pruppacher and Klett 2010). The nucleation rate depends upon the degree of subcooling, relative humidity, and the wetting properties of the surface. Once a water droplet has nucleated on the surface, it begins to grow due to the formation of concentration gradient of water molecules around the droplet. This is followed by droplet coalescence.

The atmosphere contains a large quantity of moisture in the form of water vapor. Dew point is the temperature to which air must be cooled to become saturated with water vapor. When further cooled, the airborne water vapor will condense to form water droplets or dew. At a temperature lower than the dew point, the saturated vapor pressure of water in the ambient air decreases (Alduchov and Eskridge 1996), which leads to water condensation and forms dew (Pruppacher and Klett 2010; Moran et al. 2018). It should be noted that the dew point temperature should be above the freezing point (0 °C) because dew may freeze as the temperature descends below 0 °C, and is called white dew.

The ambient temperature in the desert during nights can be low, as low as about 0–4 °C, lower than the dew point in most environments. The moisture from dew is generally available throughout the year with a larger amount available during dry summer months when living species are under highest stress when rainfall is small (Malek et al. 1999; Mares 1999).

Malek et al. (1999) measured moisture contributed by dew in northeastern Nevada and central Utah in the U.S. Based on data collected for precipitation and dew moisture contribution for northeastern Nevada with area of about 1113 km^2, between June 10, 1993 and Sept. 30, 1994, they reported that annual precipitation was about 131 mm year^{-1} and dew contribution was about 14 mm year^{-1}. Mares (1999) reported that dew contribution can range from as much as 10 L m^{-2} (10 mm) year^{-1} in cold climates to as much as 75 L m^{-2} (75 mm) year^{-1} in sub-humid warm areas. In some desert areas, especially semiarid and Mediterranean habitats, dew may contribute a significant percentage of the total precipitation that is available in the area during the year for plants and animals.

Dew can be a significant source of water for plants and animals in many of the world's deserts, particularly during the hot summer months when rainfall is small (Monteith 1963; Baier 1996; Kidron et al. 2002; Kidron 2005). In Namibia and Chile, dew is among the primary sources of water. Hill et al. (2015) measured water from dew in several plants native to the Negev desert in Israel. They reported that these plants received on the order of half of their water from dew. Dew is particularly a source of water for insects and small animals. Examples are the *Diacamma rugosum*, a common ant in India that acquires a substantial fraction of water requirements from dew, and the snail *Trichoidea seetzenii* which has been found to use dew as a source of water (Agam and Berliner 2006).

It should be noted, in general, that water collection from condensation on an annual basis is believed to be lower than that collected from the fog by plants and animals.

2.2 Desert Plants and Water Harvesting Mechanisms

To survive in deserts, plants have adapted to the extremes of heat and aridity by using both physical, chemical and behavioral mechanisms (Anonymous 1996; Gurera and Bhushan 2020). Plants that have adapted by altering their physical structure are called xerophytes. Xerophytes, such as cacti, usually have special means of storing and conserving water. They often have few, small or narrow leaves or no leaves, which reduces transpiration. Phreatophytes are plants that have adapted to arid environments by growing extremely long roots, allowing them to acquire moisture at or near the water table. Poikilohydric desert plants live for several years, often survive by remaining dormant during dry periods of the year, then springing to life when water becomes available. Other desert plants, using behavioral adaptations, have developed a lifestyle in conformance with the seasons of greatest moisture and/or coolest temperatures. These type of plants live for only a season.

Figure 2.11 presents the classification of desert plants, with examples for each classification (Anonymous 1996; Gurera and Bhushan 2020). Plants, in general, can broadly be categorized into four classes—wildflowers, cacti and other succulents, trees and shrubs, and grasses, mosses and lichens. Figure 2.12 presents a montage of selected plants from each class (Anonymous 1996; Gurera and Bhushan 2020).

Desert plants

Wildflowers

Alkali Mariposa Lily, Alkali Phacelia, Antelope Bush, Apache Plume, Arizona Poppy, Asian Mustard, Aven Nelson's Phacelia, Beautiful Centaury (Canchalagua), Big Sagebrush, Bigelow's Monkeyflower, Bitter Root, Blacktack Phacelia, Bladderpod, Blazing Star, Blue Phacelia, Blue Eyed Grass, Blue Flax, Booth's Sun Cup, Bristly Langloisia, Brittlebush, Broad-Leaf Aliciella, Broad-Leaf Gilia, Brown-Eyed Evening Primrose, Burrobush, Burro Weed, Button Brittlebush, Button Encelia, California Buckeye, California Fuschia, California Milkweed, California Poppy, Caltha-leafed Phacelia, Canaigre Dock, Canterbury Bells, Caterpillar Phacelia, Cave Primrose, Chaparral Mallow, Charlotte's Phacelia, Charming Centaury (Canchalagua), Checker Fiddleneck, Cheesebush, Burrobrush, Chia, Chinchweed, Chinese Parsley, Chocolate Drops, Chuparosa, Cliffrose, Clustered Broom-Rape, Cold-desert Phlox, Common Muilla, Common Yellow Monkeyflower, Cooper's Dogweed, Coulter's Jewelflower, Coyote Melon, Cream Cups, Crimson Columbine, Crowned Muilla, Daisy Desertstar, Datura, Death Valley Monkeyflower, Desert Bitterbrush, Desert Bluebells, Desert Broom Rape, Desert Candle, Desert Chicory, Desert Christmas Tree, Desert Dandelion, Desert Five-Spot, Desert Globemallow, Desert Holly, Desert Indian Paintbrush, Desert Indian Tobacco, Desert Larkspur, Desert Lily, Desert Lupine, Desert Marigold, Desert Mariposa Lily, Desert Milkweed, Desert Peach, Desert Primrose, Desert Paintbrush, Desert Penstemon, Desert Pincushion, Desert Poppy, Desert Purple Mat, Desert Sage, Desert Spring Parsley, Desert Star, Desert Star-Vine, Desert Sunflower, Desert Willow, Devil's Lettuce, Dogbane, Downy Dalea, Devil's Claw, Dune Evening Primrose, Dunes Sunflower, Eaton Firecracker, Eaton Penstemon, Fairy Duster, Fiddleneck, Flaxleaf Monardella, Freckled Milkvetch, Fremont's Monkeyflower, Fremont's Phacelia, Filaree Storksbill, Fringed Amaranth, Fringed Onion, Ghost Flower, Golden Desert Snapdragon, Gooding's Verbena, Gray Ball Sage, Gravel Ghost, Great Basin Woollystar, Ground Cherry, Hairy Wild Cabbage, Heartleaf Sun Cup, Hooker's Evening Primrose, Hop Sage, Hummingbird Trumpet, Indian Paintbrush, Indian Tobacco, Jewelflower, Jimson Weed, Kelso Creek Monkeyflower, Lacy Phacelia, Layne Locoweed, Lobeleaf Groundsel, Loco Weed, Mariposa Lily, Martin's Paintbrush, Mexican Whorled Milkweed, Miner's Lettuce, Mojave Aster, Mojave Beardtongue, Mojave Hole-in-the-Sand Plant, Mojave Indian Paintbrush, Mojave Indigo Bush, Mojave Monkeyflower, Mojave Prickly Poppy, Mojave Sun Cup, Narrow Leaf Milkweed, Nevada Onion, Notch-leaved Phacelia, Owl's Clover (Castilleja exserta ssp. exserta), Pale-Yellow-Sun-Cup, Palmer's Penstemon, Panamint Daisy, Panamint Mariposa Lily, Paperbag Bush, Parish's Monkeyflower, Parry's Nolina (Parry's Beargrass), Pedicellate Phacelia, Pink Phlox, Plantago patagonica, Plummer's Mariposa Lily, Poodle Dog Bush, Popcorn Flower, Prickly Poppies, Prince's Rock Cress, Purple Mat, Purplemat, Purple Owl's Clover (Castilleja exserta ssp. venusta), Purple Sand Food, Pygmy Poppy, Rattlesnake Weed, Rattleweed, Red Maids, Rock Cress, Rock Daisy, Rose Heath, Round Leafed Phacelia, Roundleaf Phacelia, Rush Milkweed, San Bernardino Mountains Liveforever, Sand Blazing Star, Sand Pygmyweed, Sand Verbena, Sapphire Woollystar, Scale Bud, Scaly-stemmed Sand Plant, Scarlet Bugler, Scorpionweed, Sea Muilla, Shockley Evening Primrose, Showy Four O'Clock, Showy Milkweed, Silky Dalea, Skeleton Milkweed, Soft Prairie Clover, Spanish Needle, Spectacle Pod, Specter Phacelia, Suksdorf's Monkeyflower, Tree Poppy, Tree Tobacco, Turtleback, Velvet Turtleback, Twining Snapdragon, Washoe Phacelia, Western Desert Penstemon, Western Wallflower, Western Forget-Me-Not, White Bear Poppy, White Bursage, White Fiesta Flower, White Mallow, White Sage, White Wooly Daisy, Whitemargin Beardtongue, White-margined Beardtongue, Wild Heliotrope, Wild Rhubarb, Winding Mariposa Lily, Wishbone Bush, Woody Bottle Washer, Woolly Daisy, Woolly Marigold, Yellow Beeplant, Yellow Cups, Yellow Desert Evening Primrose, Yellow Nightshade, Yellow Pincushion, Yerba Mansa

Cacti and other Succulents

Barrel Cactus, Beavertail Cactus, Century Plant, Chain Fruit Cholla, Cholla Cactus, Claret Cup Cactus, Desert Christmas Cactus, Datil Yucca, Fishhook Cactus, Hedgehog Cactus, Kingcup Cactus, Mojave Mound Cactus, Mojave Yucca, Night-Blooming Cereus, Organ Pipe Cactus, Prickly Pear Cactus, Saguaro Cactus, Senita Cactus, Soaptree Yucca

Trees and Shrubs	Grasses, Mosses and Lichens
Bastard Toadflax, Crucifixion Thorn, California Fan Palm, Cottonwood, Creosote Bush, Desert Lavender, Desert Willow, Elephant Tree, Greasewood, Joshua Tree, Juniper, Mormon Tea, Ocotillo, Mesquite Tree, Palo Verde Tree, Poison Ivy, Ponderosa Pine, Four-Wing Saltbush, Showy Milkweed, Smoke Tree, Winterfat	Bushman Grass, Mitten Steppe Screw Moss, Lichen

(Adapted from https://www.desertusa.com/flora.html)

Fig. 2.11 Classification of desert plants with examples for each class (adapted from Gurera and Bhushan 2020)

2.2.1 Wildflowers

Several adaptations have enabled wildflowers to thrive in the heat and dryness of their habitats. Some plants will only blossom on the rare occasions when water appears in the desert, and remain dormant for the rest of the year. Others only grow

Fig. 2.12 Montage of selected desert plants with their class for each class (adapted from Gurera and Bhushan 2020)

during rainy seasons and have short lives, such as the desert sand verbena, which grows and blooms with bright purple flowers after rainfalls. Its seeds can remain in the ground for months or years before growing after the next rainy season. Some wildflowers which include desert lily, desert lupine, desert marigold, fairy duster, twist flower and larkspur, capitalize on fog and dew for their water needs (Von Hase et al. 2006; Gurera and Bhushan 2020). The droplets are deposited on the leaves, which are slowly adsorbed.

2.2.2 Cacti and Other Succulents

Succulents are a type of plants with thick, fleshy and swollen parts that are adapted to store water and to minimize water loss (Ogburn and Edwards 2009; Gurera and Bhushan 2020). Succulents, in general, have evolved a number of strategies for holding onto this water. They tend to have a thick waxy coating, which helps to seal in moisture. They are commonly found in arid regions.

While cacti are by definition succulents, they are often referred to separately from other succulents. These species use fog as a supplemental source of water (Mooney et al. 1977). A cactus specie, *Copiapoa haseltoniana*, native to the Atacama Desert, is known to use the run off from nightly fog events to survive in what is the driest non-polar desert in the world (Vesilind 2003). *Gymnocalycium baldianum*, a cactus specie in neighboring Argentina, is known to collect and transport water through microcapillaries (Liu et al. 2015). A specie, *Opuntia microdasys* (angle's wings, bunny ears cactus or polka dot cactus), endemic to Mexico, features conical spines with small barbs atop that help in the collection of water (Ju et al. 2012; Brown and Bhushan 2016; Bhushan 2018, 2019, 2020; Gurera and Bhushan 2020). *Ferocactus latispinus* specie also consists of conical spines with small barbs (Bhushan 2018, 2020). Water droplets collect on the tips of the small barbs and once they reach critical size, they move onto the conical spine. On the spine, the droplets move towards the base due to the curvature gradient, providing the Laplace pressure gradient. Once at the base of the spine, the plant absorbs the water. Laplace pressure gradient is large enough that water droplets can defy gravity and climb upward (Bhushan 2018).

Another common succulent found in deserts is *Trianthema hereroensis* (*Aizoaceae*) which is prevalent in the coastal desert. This plant rapidly absorbs fog water through its leaves, apparently to supplement the water obtained through its roots. It only grows as far inland as the fog regularly penetrates. As a result of this relatively dependable water supply, it flowers and produces seeds throughout the year and is thus an important source of food and shelter for many dune animals (Seely et al. 1977).

2.2.3 Trees and Shrubs

Desert trees such as the Joshua tree collect water on their leaves and branches from rain, fog, and dew, and store it in its trunk and leaves. Large Joshua trees are about 10 m tall and their widespread roots are about 1 m deep. The collected water coalesces and drips. However, trees do not consist of any water harvesting mechanisms.

Shrubs such as the Creosote bush have fine hairs on the leaves of the plant that intercept fog droplets where they coalesce and grow before dropping down into the plant structure when they become too heavy, and eventually reach the roots (Bhushan 2018, 2020).

2.2.4   Grasses, Mosses and Lichens

A grass endemic to the Namib Desert, *Stipagrostis sabulicola* (Bushman grass), collects water from fog (Brown and Bhushan 2016; Bhushan 2018, 2019, 2020; Gurera and Bhushan 2020). Water droplets collect on the leaf before coalescing and running down towards the base of the plant (Louw and Seely 1980; Ebner et al. 2011). The leaves feature longitudinal ridges, which facilitate water flow (Roth-Nebelsick et al. 2012; Brown and Bhushan 2016). Another type of grass, *Setaria viridis*, is also found to harvest water from fog by directionally transporting water droplets using its grooves and conical shape (Xue et al. 2014).

Syntrichia caninervis (Mitten Steppe Screw Moss) is one of the most abundant desert mosses in the world and thrives in extreme environments with limited water resources (dew, fog, snow and rain) (Koch et al. 2008; Pan et al. 2016; Bhushan 2020; Gurera and Bhushan 2020). *S. caninervis* has a unique adaptation. It uses a tiny hair (awn) on the end of each leaf to collect water, in addition to that collected by the leaves themselves. Water droplets collect on the awn's barbs which serve as collection depots. When droplets become large enough, they move down through micro-/nanogrooves along the length of awn to the base of the leaf (Gurera and Bhushan 2020).

Lichens comprise a fungus living in a symbiotic relationship with an alga or cyanobacterium (or both in some instances) (Mares 1999; Gurera and Bhushan 2020). Examples include *Teloschistes capensis*, *Alectoria*, *Santessonia hereroensis*, *Caloplaca elegantissima*, *Xanthomaculina hottentotta*, and *Xanthomaculina convolute*. Lichens are abundant in both hot and cold deserts. They form crusts on exposed rock and through the release of chemicals, dissolve the substrate to form soil. During periods of extreme drought, lichens become dry and appear to be dead, but with rain they absorb water and become green again. Lichens are poikilohydric, which means they have the capacity to tolerate dehydration to low cell or tissue water content and to recover from it without physiological damage. They do not have water storing tissues or a waxy cuticle (Mares 1999; Gurera and Bhushan 2020).

2.2.5   Summary of Water Harvesting Mechanisms

A summary of water harvesting mechanisms used by various desert plants from fog and dew is presented in Table 2.5 (Gurera and Bhushan 2020). Figure 2.13 presents schematics of water harvesting mechanisms used by selected plants (Gurera and Bhushan 2020). The selected plants are desert lily (wildflower), bunny ears cactus (cactus), creosote bush (shrub), and Namib grass (grass) (Gurera and Bhushan 2020).

Table 2.5   Summary of water harvesting mechanisms by various desert plants from fog and dew (adapted from Gurera and Bhushan 2020)

Plant class	Species	Surface structures or chemistry	Harvesting mechanism	References
Wildflower	*Tripteris oppositifolium (Asteraceae) Lampranthus hoerleinianus (Aizoaceae) Grielum grandiflorum (Neuradaceae) Oxalis eckloniana (Oxalidaceae)*	Rough surface	Hydrophilic leaves and adsorption	Von Hase et al. (2006)
Cactus (succulents)	*Opuntia microdasys*[a]	Conical geometry and grooves	Water droplets get deposited on the spine, which coalesce, grow and move down onto the spine, towards the base, due to Laplace pressure gradient where they are absorbed	Mooney et al. (1977), Ju et al. (2012), and Bhushan (2018, 2019)
	Ferocactus latispinus[a]			Bhushan (2020)
	Copiapoa haseltoniana[a]			Mooney et al. (1977) and Nobel (2003)
	Discocactus horstii[a] *Turbinicarpus schmiedickeanus klinkerianus*[a] *Mammillaria theresae*[a]			Schill et al. (1973) and Nobel (2003)
	Eulychnias[a]			Yetman (2007)
	Copiapoa cinerea var. haseltoniana Ferocactus wislizenii Mammillaria columbiana subsp. yucatanensis Parodia mammulosa			Malik et al. (2015)
Other succulents	*Trianthema hereroensis (Aizooceae)*		Absorb water through leaves and quickly translocation to the roots	Seely et al. (1977) and Van Damme (1991)

(continued)

Table 2.5 (continued)

Plant class	Species	Surface structures or chemistry	Harvesting mechanism	References
Shrub	*Larrea tridentata* (Creosote bush)		Water droplets grow on tiny hairs before dropping down further into the plant structure when they get too heavy and eventually reaching the roots	Harris (2003)
	Psorothamnus arborescens (Mojave Indigo bush)			Bhushan (2020)
	Arthraerua leubnitziae (Pencil bush)			Van Damme (1991)
Grass (Namib)[a]	*Stipagrostis sabulicola*	Grooves	Water droplets are channeled down the hydrophilic leaves towards the base of the plant and eventually reaching the roots	Seely (1979), Van Damme (1991), Ebner et al. (2011), Norgaard et al. (2012), and Roth-Nebelsick et al. (2012)
Moss	*Syntrichia caninervis*	Grooves	Water droplets collect on barbs which serve as collection depots. When droplets become large enough, they move down to micro-/nanogrooves along the length of awn to its base onto the leaf	Koch et al. (2008) and Pan et al. (2016)
Lichens	*Teloschistes capensis Alectoria Santessonia hereroensis Caloplaca elegantissima Xanthomaculina hottentotta Xanthomaculina convoluta*		Absorb	Lange et al. (1991)
	Xanthoria elegans Brodoa atrofusca Umbilicaria cylindrica			Reiter et al. (2008)

[a]Adapted from Malik et al. (2014)

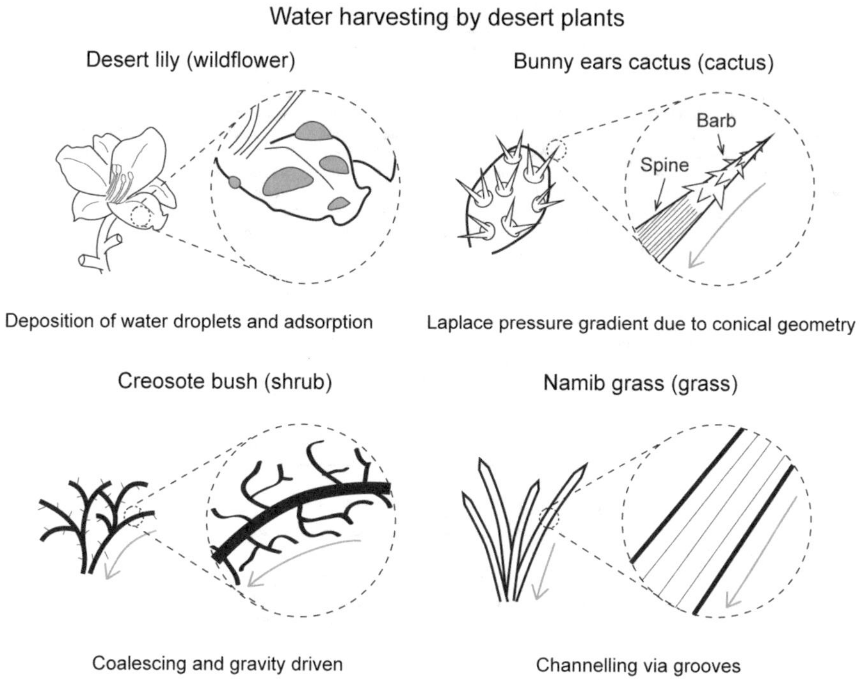

Water harvesting by desert plants

Desert lily (wildflower)   Bunny ears cactus (cactus)

Deposition of water droplets and adsorption   Laplace pressure gradient due to conical geometry

Creosote bush (shrub)   Namib grass (grass)

Coalescing and gravity driven   Channelling via grooves

Fig. 2.13 Schematics of water harvesting mechanism of selected desert plants (adapted from Gurera and Bhushan 2020)

2.3   Desert Animals and Water Harvesting Mechanisms

To survive in deserts, animals use behavioral, physiological, and anatomical adaptations (Harris 2003). The common adaptations include feeding on other animals and drinking from water sources such as rivers and lakes for fulfilling water needs. Another is by seeking cool areas (shade, soil, rocks, caves, mines, canyons, burrows, and higher elevations) during the hottest time of the day to avoid any water loss due to evaporation, perspiration and breathing. Another is being nocturnal to harvest water from fog and dew in nights because temperatures are low and humidity is high. In this section, water harvesting mechanisms used by various desert animals are discussed (Gurera and Bhushan 2020).

Figure 2.14 presents the classification of the desert animals, with examples for each classification (Gurera and Bhushan 2020). These include vertebrates and invertebrates. Vertebrates can broadly be categorized into five classes: mammals, birds, reptiles, fishes, and amphibians (Anonymous 1996). Figures 2.15 and 2.16 present montages of selected vertebrates and invertebrates, respectively (Anonymous 1996; Gurera and Bhushan 2020).

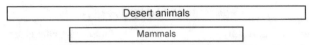

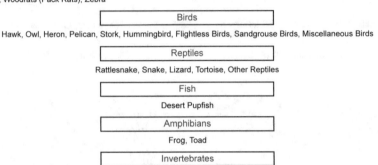

Desert animals

Mammals

African Wild Dog, Arabian Wildcat, Armadillo, Ankole Longhorn Cattle, Badger, Bats, Bighorn Sheep, Bear (Black), Bison (Plains Bison), Bobcat, Buffalo (Plains Bison), Bison - Beefalo, Burros, Cairo Spiny Mouse, Camels, Cheetah, Chipmunk, Panamint, Coati, White-Nosed, Collared Peccary, Cottontail Rabbit, Cougar, Coyote, Dama Gazelle, Deer - Mule, Deer - White-Tailed, Desert Shrew, Dingo, Dinosaurs, Dolphin , Elephant, Elk, Ferret, Finback Whale, Fox (Gray), Giraffe, Goat, Gopher, Ground Sloth, Shasta (extinct), Gray Fox, Hippopotamus, Jaguars, Jackal, Javelina, Jerboa, Wild Horses, Kangaroo Rats, Kangaroo, Red, Kit Fox, Lion, LLama, Long-Tailed Weasel, Mountain Lion, Mule Deer, Otter - River, Porcupine, Prairie Dogs, Pronghorn, Przewalski's Horse, Rabbits, Raccoon, Red Kangaroo, Rhinoceros, Ringtailed Cat, River Otter, Sea Lion - California, Shasta Ground Sloth (extinct), Skunk, Snow Leopard, Spiny Mouse, Spotted Hyena, Squirrel, Weasel, Long-Tailed, Whale - Finback, White Tail Deer, Wild Burros, Wild Horses, Wolves, California Wolf Center, Woodrats (Pack Rats), Zebra

Birds

Hawk, Owl, Heron, Pelican, Stork, Hummingbird, Flightless Birds, Sandgrouse Birds, Miscellaneous Birds

Reptiles

Rattlesnake, Snake, Lizard, Tortoise, Other Reptiles

Fish

Desert Pupfish

Amphibians

Frog, Toad

Invertebrates

Ant, Bee, Beetle, Butterfly, Flat bug, Moth, Scorpions, Spider, Spider Web, Flying Insect, Fairy Shrimp, Kanab Ambersnail, Wharf Roaches

(Adapted from https://www.desertusa.com/animals.html)

Fig. 2.14 Classification of desert animals with examples for each class (adapted from Gurera and Bhushan 2020)

2.3.1 Mammals

The mammals are the class of vertebrate animals primarily characterized by the presence of mammary glands in the female which produce milk for the nourishment of young, the presence of hair or fur, and which have endothermic or "warm blooded" bodies. Large mammals survive on rivers and lakes for their water needs. Some mammals also use water harvesting mechanisms. For example, in the case of desert elephants, *Loxodonta Africana* hold water on their skin which provides many benefits (Lillywhite and Stein 1987; Anonymous 2016; Martins et al. 2018; Gurera and Bhushan 2020). They have an intricate network of crevices in their skin surface. These micrometer-wide channels enhance the effectiveness of thermal regulation (by water retention) as well as protection against parasites and intense solar radiation by mud adherence. This fine pattern of channels allows spreading and retention of 5–10 times more water on the elephant skin than on a flat surface, impeding dehydration and improving thermal regulation over a longer period of time.

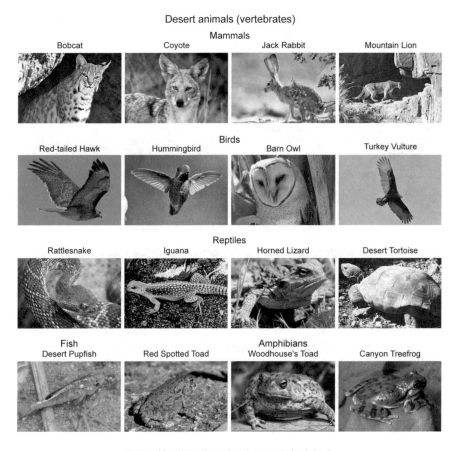

Fig. 2.15   Montage of selected desert animals (vertebrates) for each class (adapted from Gurera and Bhushan 2020)

2.3.2   Birds

Birds, also known as Aves, are a group of endothermic vertebrates, characterized by feathers, toothless beaked jaws, the laying of hard-shelled eggs, a high metabolic rate, a four-chambered heart, and a strong yet lightweight skeleton (Anonymous 1995; Harris 2003). Most birds can fly. Not all flying animals are birds, and not all birds can fly. Examples of flying birds include sandgrouse and elf owls. Male sandgrouse, to satisfy the thirst of newly hatched chicks, bring water back to the nest by carrying it in their feathers (Cade and Maclean 1967). In the cool of the

Desert animals (invertebrates)

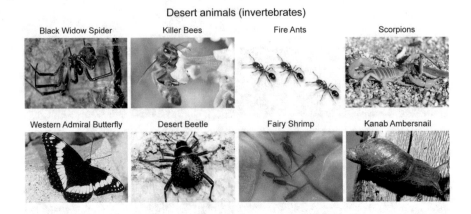

(Adapted from https://www.desertusa.com/animals.html)

Fig. 2.16 Montage of selected desert animals (invertebrates) (adapted from Gurera and Bhushan 2020)

desert morning, the male flies up to 30 km to a shallow water hole, then wades in up to his belly. The water is collected by "rocking." The bird shifts its body side to side and repeatedly shakes the belly feathers in the water. Fill-up can take as long as 15 min. Due to coiled hairlike extensions on the feathers of the underparts, a sandgrouse can soak up and transport on the order of 25 mL of water, about two tablespoons. Once the male has flown back across the desert with his life-giving cargo, the sandgrouse chicks crowd around him and use their bills to consume that water (Gurera and Bhushan 2020).

2.3.3 *Reptiles*

Reptiles are animals in the class *Reptilia*. Three of the commonly known desert reptiles are snakes, lizards, and tortoise. Reptiles mostly harvest their water via their food and/or usually just lick and drink water from stones or vegetation (Joel et al. 2017). Reptiles avoid water loss by being inactive during the hottest times of a day. Their waterproof skin helps them to minimize water loss through evaporation. They also simultaneously withhold and/or drink any water available.

2.3.3.1   Snakes

Water harvesting snakes emerge from their dens or any potential roofing during rain showers and coil up in the open. By coiling up, they bring their body loops into close contact with each other and water is restrained in the formed shallows between the loops (Joel et al. 2017; Bhushan 2020). Some species also exhibit a dorsoventral flattening of the body, probably to increase the exposed surface area. With their noses continuously in contact with the shallows and the head moving contrary to scales' overlap, the collected water is ingested (Gurera and Bhushan 2020).

2.3.3.2   Lizards

Moloch horridus is a lizard specie native to arid regions in western and southern Australia (Bentley and Blumer 1962). Water droplets deposited on the body spread out over the skin before reaching the mouth. Water movement occurs due to capillary action along open channels in the skin. The lizard adopts a posture where the head is depressed and the hindquarters are elevated which helps guide the water to the mouth as reported for *Phrynocephalus helioscopus*, a lizard specie native to arid regions in Asia (Schwenk and Greene 1987; Bhushan 2018).

Similarly, in the case of *Phrynosoma cornutum* or the Texas horned lizard, water placed on the skin flows preferentially towards the mouth, with capillary forces (Comanns et al. 2015). A network of capillary channels exist with a narrowing of individual channels in the direction of the mouth (longitudinal). The narrowing of the channels results in favorable water transport in that direction due to the curvature of the liquid–air interface. In the backward direction, liquid flow is stopped as the channel widens and pressure would need to be applied to force the liquid in that direction (Brown and Bhushan 2016; Bhushan 2018).

2.3.3.3   Tortoises

A posture like lizards is adopted by water harvesting tortoises (Joel et al. 2017). However, they press the front limbs against their head, probably to lead the water from their shell towards their snout. Any flattening of their body is, apparently, inhibited by their rigid protection (Gurera and Bhushan 2020).

2.3.4   Fish

Fish species in desert rivers are adapted to live in highly fluctuating ecosystems (Unmack 2001; Clavero et al. 2015). One of the known species of desert fish is *Cyprinodon macularius* (desert pupfish) (Anonymous 1996). Some of these fishes can tolerate temperatures between approximately 4 and 45 °C and salinities ranging from 0 to 70 parts per thousand, exceeding the tolerances of virtually all other fresh water fish. The desert pupfish can also survive dissolved-oxygen concentrations as low as 0.13 ppm. However, they do not consist of water harvesting mechanisms (Gurera and Bhushan 2020).

2.3.5   Amphibians

Amphibians are ectothermic, tetrapod vertebrates of the class *Amphibia*. Examples include toads and frogs. Amphibians have developed morphological and behavioral adaptations to maintain the hydric balance (Lillywhite and Licht 1974; Harris 2003; Comanns 2018; Gurera and Bhushan 2020). Amphibians typically absorb water through the skin. For example, surrounded by damp soil in a deep burrow, the toad acts like a sponge and draws water into itself from the soil. Some hylid toads of the genus *Anaxyrus* (e.g. *A. boreas*, *A. woodhousii*, *A. punctatus*) have granular skin which contains numerous grooves in which capillary forces suck water from the surface. A granular ventral skin absorbs more efficiently than a smooth skin. Their accumulated water is transported even to the dorsal body parts, most likely to maximize effective wetting of the skin to enlarge the area for water uptake and to prevent dehydration of the epidermis (Gurera and Bhushan 2020).

Some hylid frogs, such as *Litoria caerulea* (Australian green treefrog) and *Phyllomedusa sauvagii* (waxy monkey tree frog) survive on water droplets from condensation of water vapor (Toledo and Jared 1993; Tracy et al. 2011). The required thermal gradient is achieved by the ectothermic properties of the species and temporal changes in the microhabitat. As ectotherms, the frogs cool down in the open environment. However, when they enter warm and humid tree holes, the temperature difference leads to condensation on the colder body surfaces of the frog. Condensation rates up to 3 mg cm^{-2} on the body surface at a temperature differential of 15 °C and duration of 20 min have been reported (Tracy et al. 2011). Such a rate is higher than the water loss by evaporation.

2.3.6   Invertebrates

Examples of invertebrates include insects and crustaceans. Many deserts have more insect species than all other animal groups combined. Insects common to deserts include ants, spiders, bees, beetles, grasshoppers, butterflies, flat bugs, moths and scorpions. They have developed many adaptations and behaviors to help them survive heat, drought and predators in the desert. Crustaceans are a group of aquatic animals. Common crustaceans in deserts include fairy shrimp, snails and wharf roaches (Gurera and Bhushan 2020).

2.3.6.1 Beetles

Beetles are commonly found in very arid Namib desert. *Stenocara gracilipes* and *Onymacris unguicularis* beetles are native to this region. The thick exoskeletons of beetles help minimize water loss, and the cavities beneath their forewings trap moisture. The beetles survive as a result of water harvesting from fog. Beetles lower their heads while oriented into the wind. Water trickles down their body and into the mouth (Hamilton and Seely 1976). The back of the beetle comprises a random array of about 0.5 mm diameter bumps, spaced 0.5–1.5 mm apart. The bumps are smooth and hydrophilic, while the surrounding area is covered with microstructured wax which is hydrophobic. Water from the fog is observed to land on the bumps and droplets begin to grow. The droplet continues to grow (on the order of 5 mm) until the weight of the droplet overcomes the capillary force and the droplet detaches and rolls down the tilted beetle's back to its mouth (Parker and Lawrence 2001; Brown and Bhushan 2016; Bhushan 2018, 2019).

Another beetle, similar to the Namib beetle, known as the flower beetle, has a similar water harvesting mechanism (Godeau et al. 2018). One of the flower beetle specie, *G. orientalis* is a large insect and can grow up to 110 mm long. It is well known for its distinctive black and white regions, which are hydrophobic and hydrophilic, respectively. The white regions are covered with horizontally aligned nanohairs as well as vertically aligned microhairs. Surfaces of microhairs contain nanogrooves. The nanogrooves may preferentially direct the movement of water droplets for water harvesting.

2.3.6.2 Flat Bugs

The South American flat bug species, *Dysodius lunatus* and *Dysodius magnus*, collect water for camouflage, in order to reduce their surface reflectivity, rather than for the purpose of rehydration (Silberglied and Aiello 1980; Hischen et al. 2017; Gurera and Bhushan 2020). Immediate spreading of water droplets is facilitated by chemical and structural properties of the integument. Unlike most other insects, the cuticle of these bugs is covered by a hydrophilic wax layer imparted by the amphiphilic component erucamide (Hischen et al. 2017). In the flat bug species,

D. lunatus and *D. magnus*, pillar-like surface structures provide the hydrophilic wetting properties and spreading of water. Spreading on the surface is slower than within integumental channels, but energetically favorable. Hence, a passive spreading of water over the body surface is enabled (Hischen et al. 2017).

2.3.6.3 Crustaceans

Crustaceans are a group of aquatic animals that include fairy shrimps, snails, and wharf roaches (Anonymous 1996; Harris 2003; Gurera and Bhushan 2020). Fairy shrimp are tiny fresh water crustaceans related to lobsters, shrimps and crabs. Fairy shrimp spend their entire lives in ephemeral pools, often located in very remote areas. Snails use their mucus to stick their shells to a hard substrate such as the underside of a shady rock. The mucus dries, creating a waterproof seal around the shell and keeping the snail moist inside for months (Yom-Tov 1971; Degen et al. 1992).

Wharf roaches, such as *Ligia exotica* and *Ligia oceanica*, found in coastal deserts, passively collect water from wet surfaces of coastal habitat (Hoese 1981; Horiguchi et al. 2007). Water is transported in open structures of the cuticle of the legs, which act as capillaries. The hair- and paddle-like microstructures on their two adjacent legs collect and transport the adhered water. The water is transported further along the swimming limbs (pleopods) and to the hindgut, near the anus, for uptake by absorption (Horiguchi et al. 2007; Ishii et al. 2013). Collected water also establishes a water film on the integument and evaporation provides thermoregulation (Hoese 1981).

2.3.7   Summary of Water Harvesting Mechanisms

A summary of water harvesting mechanisms used by various desert animals is presented in Tables 2.6 and 2.7 for vertebrates and invertebrates, respectively (Gurera and Bhushan 2020). Figure 2.17a and b presents schematics of water harvesting mechanisms used by selected vertebrates and invertebrates, respectively (Gurera and Bhushan 2020). The selected vertebrates are elephant (mammals), sandgrouse (birds), lizard (reptiles), and toad (amphibians). The selected invertebrate is Namib beetle (insects).

Table 2.6 Summary of water harvesting mechanisms used by various desert animals (vertebrates) (adapted from Comanns 2018; Gurera and Bhushan 2020)

Animal class	Species	Water source	Surface structures or chemistry	Harvesting mechanism	References
Mammals (elephants)	*Loxodonta africana*	Open water/lakes	Ridges/grooves	Skin-wetting properties, capillarity	Lillywhite and Stein (1987)
Birds (sandgrouse)	*Pterocles bicinctus*	Open water/lakes	Hairy feather structure	Storage between feathers, direct drinking by chicks	Cade and Maclean (1967)
	Pterocles namaqua				Joubert and Maclean (1973)
Reptiles (snakes)	*Crotalus atrox*	Rain		Accumulation of collected water	Repp and Schuett (2008)
	Crotalus mitchellii pyrrhus				Glaudas (2009)
	Crotalus viridis concolor				Ashton and Johnson (1998)
	Crotalus s. scutulatus				Cardwell (2006)
	Bothrops moojeni				Andrade and Abe (2000)
	Bitis peringueyi	Rain, fog			Louw (1972) and Robinson and Hughes (1978)
Reptiles (lizards)	*Phrynosoma cornutum*	Moist substrate, rain	Honeycomb-like micro-structure, channels between the scales	Transport in channels between scales from all body parts to mouth for drinking	Sherbrooke (1990, 2004) and Comanns et al. (2015)
	Phrynosoma modestum				Sherbrooke (2002)
	Phrynosoma platyrhinos				Pianka and Parker (1975) and Peterson (1998)

(continued)

Table 2.6 (continued)

Animal class	Species	Water source	Surface structures or chemistry	Harvesting mechanism	References
	Phrynocephalus arabicus				Comanns et al. (2011)
	Phrynocephalus helioscopus				Schwenk and Greene (1987)
	Phrynocephalus horvathi				Yenmiş et al. (2016)
	Trapelus flavimaculatus *Trapelus pallidus* *Trapelus mutabilis*				Veselý and Modrý (2002)
	Moloch horridus	Rain	Channels between the scales		Bentley and Blumer (1962), Gans et al. (1982), Sherbrooke (1993), and Withers (1993)
	Uromastyx spinipes				Ditmars (1936)
	Pogona vitticeps				Fitzgerald (1983)
	Aporosaura anchietae	Fog		Fog basking	Louw (1972)
Reptiles (tortoises)	*Psammobates tentorius trimeni* *Kinixys homeana* *Homopus areolatus*	Rain	Large ridges of carapace	Gravity-facilitated transport on surface to mouth for drinking	Auffenberg (1963)

(continued)

Table 2.6 (continued)

Animal class	Species	Water source	Surface structures or chemistry	Harvesting mechanism	References
Amphibians (toads)	*Anaxyrus boreas*	Moist substrate	Ridges, channels	Skin wetting and capillary transport to replenish evaporative loss	Fair (1970), Lillywhite and Licht (1974), and Toledo and Jared (1993)
	Anaxyrus woodhousii				Lillywhite and Licht (1974)
	Anaxyrus punctatus				McClanahan and Baldwin (1969) and Fair (1970)
Amphibians (treefrogs)	*Phyllomedusa sauvagii*	Humidity, dew	Hygroscopic secretion, slightly granular skin	Transcutaneous uptake	Shoemaker et al. (1972) and Toledo and Jared (1993)
	Litoria caerulea				Toledo and Jared (1993) and Tracy et al. (2011)

Table 2.7 Summary of water harvesting mechanisms by various desert animals (invertebrates) (adapted from Comanns 2018; Gurera and Bhushan 2020)

Animal class	Species	Water source	Surface structures or chemistry	Collection mechanism	References
Insects (beetles)	*Onymacris unguicularis*	Fog	Hydrophobic body covered with wax-free hydrophilic bumps	Wetting properties of elytra, gravity	Hamilton and Seely (1976), Seely (1979), and Norgaard and Dacke (2010)
	Onymacris bicolor				Seely (1979) and Norgaard and Dacke (2010)
	Stenocara sp.				Parker and Lawrence (2001)
Insects (flat bugs)	*Dysodius lunatus Dysodius magnus*	Rain	Hydrophilic waxes, spine microstructures, channels	Reducing reflectivity, aiding camouflage	Hischen et al. (2017)
Crustacea (wharf roaches)	*Ligia exotica*	Moist substrate	Channels between hair-like and paddle-like protrusions	Thermoregulation or transport to hindgut for uptake	Horiguchi et al. (2007) and Ishii et al. (2013)
	Ligia oceanica				Hoese (1981)

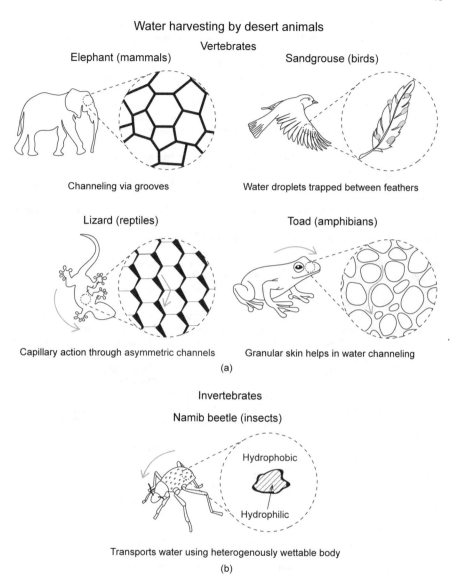

Fig. 2.17 Schematics of water harvesting mechanism of selected desert animals **a** vertebrates and **b** invertebrates (adapted from Gurera and Bhushan 2020)

References

Agam, N. and Berliner, P. R. (2006), "Dew Formation and Water Vapor Adsorption in Semi-Arid Environments – A Review," *J. Arid Environ.* **65**, 572-590.

Alduchov, O. A. and Eskridge, R. E. (1996), "Improved Magnus Form Approximation of Saturation Vapor Pressure," *J. Appl. Meteorol.* **35**, 601-609.

Allaby, M. (2006), *Deserts*, Chelsea House Publishers, New York, New York.

Andrade, D. and Abe, A. (2000), "Water Collection by the Body in a Viperid Snake, Bothrops Moojeni," *Amphibia-Reptilia* **21**, 485–492.

Anonymous (1995), "Discover and Learn about Birds– Desert Birds: How Do Birds Drink Water? Part 1," https://www.birds.com/desert-birds-how-do-birds-drink-water-part-1/.

Anonymous (1996), "DesertUSA," https://www.desertusa.com/.

Anonymous (1999), "Fog, Dew and Snow Harvesting," in *Sourcebook of Alternative Technologies for Freshwater Augumentation in Africa*, Stationery Office Books, Norwich, U.K.

Anonymous (2016), "Mammals in the Desert," *adventurepublications.net*, http://adventurepublications.net/2016/12/28/desert/.

Ashton, K. G. and Johnson, J. (1998), "Crotalus Viridis Concolor (Midget Faded Rattlesnake): Drinking from Skin," *Herpetol. Rev.* **29**, 170.

Auffenberg, W. (1963), "A Note on the Drinking Habits of Some Land Tortoises," *Anim. Behav.* **11**, 72–73.

Axelrod, D. I. (1979), *Age and Origin of Sonoran Desert Vegetation*, California Academy of Sciences, San Francisco, California.

Baier, W. (1996), "Studies on Dew Formation under Semi-arid Conditions," *Agric. Meteorol.* **3**, 103-112.

Basu, S., Agarwal, A. K., Mukhopadhyay, A., and Patel, C. (Eds.) (2018), *Droplet and Spray Transport: Paradigms and Applications*, Springer, New York.

Bentley, P. J. and Blumer, W. F. C. (1962), "Uptake of Water by the Lizard, Moloch Horridus," *Nature* **194**, 699–700.

Bhushan, B. (2018), *Biomimetics: Bioinspired Hierarchical-Structured Surfaces for Green Science and Technology*, third ed., Springer International, Cham, Switzerland.

Bhushan, B. (2019), "Bioinspired Water Collection Methods to Supplement Water Supply," *Phil. Trans. R. Soc. A* **377**, 20190119.

Bhushan, B. (2020), "Design of Water Harvesting Towers and Projections for Water Collection from Fog and Condensation," *Phil. Trans. R. Soc. A* **378**, 20190440.

Bredeson, C. (2009), *Baby Animals of the Desert*, Enslow Publishers, Berkeley Heights, New Jersey.

Brown, P. S. and Bhushan, B. (2016), "Bioinspired Materials for Water Supply and Management: Water Collection, Water Purification and Separation of Water from Oil," *Phil. Trans. R. Soc. A* **374**, 20160135.

Cade, T. J. and Maclean, G. L. (1967), "Transport of Water by Adult Sandgrouse to Their Young," *Condor* **69**, 323–343.

Cardwell, M. D. (2006), "Rain-Harvesting in a Wild Population of Crotalus s. Scutulatus (Serpentes: Viperidae)," *Herpetol. Rev.* **37**, 142–144.

Clavero, M., Esquivias, J., Qninba, A., Riesco, M., Calzada, J., Ribeiro, F., Fernández, N., and Delibes, M. (2015), "Fish Invading Deserts: Non-Native Species in Arid Moroccan Rivers," *Aquat. Conserv.: Mar. Freshw. Ecosyst.* **25**, 49–60.

Cloudsley-Thompson, J. L. and Chadwick, M. J. (1964), *Life in Deserts*, G. T. Foulis & Co., London, U. K.

Comanns, P. (2018), "Passive Water Collection with the Integument: Mechanisms and Their Biomimetic Potential," *J. Exp. Biol.* **221**, jeb153130.

Comanns, P., Effertz, C., Hischen, F., Staudt, K., Böhme, W., and Baumgartner, W. (2011), "Moisture Harvesting and Water Transport through Specialized Micro-Structures on the Integument of Lizards," *Beilstein J. Nanotechnol.* **2**, 204–214.

Comanns, P., Buchberger, G., Buchbaum, A., Baumgartner, R., Kogler, A., Bauer, S., and Baumgartner, W. (2015), "Directional, Passive Liquid Transport: the Texas Horned Lizard as a Model for a Biomimetic 'Liquid Diode'," *J. R. Soc. Interface* **12**, 20150415.

Costa, G. (1995), *Behavioural Adaptations of Desert Animals*, Springer-Verlag, Berlin, Germany.

Degen, A. A., Leeper, A., and Shachak, M. (1992), "The Effect of Slope Direction and Population Density on Water Influx in a Desert Snail, Trochoidea Seetzenii," *Funct. Ecol.* **6**, 160–166.

Ditmars, R. L. (1936), *Reptiles of the World*, The MacMillan company, New York, New York.

Ebner, M., Miranda, T., and Roth-Nebelsick, A. (2011), "Efficient Fog Harvesting by Stipagrostis Sabulicola (Namib Dune Bushman Grass)," *J. Arid. Environ.* **75**, 524–531.

Fair, J. W. (1970), "Comparative Rates of Rehydration from Soil in Two Species of Toads, Bufo Boreas and Bufo Punctatus," *Comp. Biochem. Physiol* **34**, 281–287.

Fessehaye, M., Abdul-Wahab, S. A., Savage, M. J., Kohler, T., Gherezghiher, T., and Hurni, H. (2014), "Fog-Water Collection for Community Use," *Renew. Sust. Energ. Rev.* **29**, 52–62.

Fitzgerald, M. (1983), "A Note on Water Collection by the Bearded Dragon Amphibolurus Vitticeps," *Herpetofauna* **14**, 93.

Gans, C., Merlin, R., and Blumer, W. F. C. (1982), "The Water-Collecting Mechanism of Moloch Horridus Re-Examined," *Amphibia-Reptilia* **3**, 57–64.

Glaudas, X. (2009), "Rain-Harvesting by the Southwestern Speckled Rattlesnake (Crotalus Mitchellii Pyrrhus)," *Southwest. Nat.* **54**, 518–521.

Godeau, G., Godeau, R.-P., Orange, F., Szczepanski, C. R., Guittard, F., and Darmanin, T. (2018), "Variation of Goliathus Orientalis (Moser, 1909) Elytra Nanostructurations and Their Impact on Wettability," *Biomimetics.* **3**, 6.

Greenberger, R. (2009), *Deserts–The Living Landscape*, The Rosen Publishing Group, New York, New York.

Gurera, D. and Bhushan, B. (2020), "Passive Water Harvesting by Desert Plants and Animals: Lessons from Nature," *Phil. Trans. R. Soc. A* **378**, 20190444.

Hamilton, W. J., and Seely, M. K. (1976), "Fog Basking by the Namib Desert Beetle, *Onymacris Unguicularis*," *Nature* **262**, 284–285.

Harris, N. (2003), *Atlas of the World's Deserts*, Taylor and Francis Group, New York, New York.

Herbert, F. (1965), *Dune*, Berkeley Publishing Corp., New York, New York.

Hiatt, C., Fernandez, D., and Potter, C. (2012), "Measurements of Fog Water Deposition on the California Central Coast," *Atmospheric Clim. Sci.* **02**, 525.

Hill, A. J., Dawson, T. E., Shelef, O., and Rachmilevitch, S. (2015). "The Role of Dew in Negev Desert Plants," *Oecologia*, **178**, 317–327.

Hischen, F., Reiswich, V., Kupsch, D., Mecquenem, N. D., Riedel, M., Himmelsbach, M., Weth, A., Heiss, E., Armbruster, O., Heitz, J., and Baumgartner, W. (2017), "Adaptive Camouflage: What Can Be Learned from the Wetting Behaviour of the Tropical Flat Bugs Dysodius Lunatus and Dysodius Magnus," *Biol. Open* **6**, *1209–1218*.

Hoese, B. (1981), "Morphologie und Funktion des Wasserleitungssystems der terrestrischen Isopoden (Crustacea, Isopoda, Oniscoidea)," *Zoomorphology* **98**, 135–167.

Horiguchi, H., Hironaka, M., Meyer-Rochow, V. B., and Hariyama, T. (2007), "Water Uptake via Two Pairs of Specialized Legs in Ligia Exotica (Crustacea, Isopoda)," *Biol. Bull.* **213**, 196–203.

Ishii, D., Horiguchi, H., Hirai, Y., Yabu, H., Matsuo, Y., Ijiro, K., Tsujii, K., Shimozawa, T., Hariyama, T., and Shimomura, M. (2013), "Water Transport Mechanism through Open Capillaries Analyzed by Direct Surface Modifications on Biological Surfaces," *Sci. Rep.* **3**, 3024.

Joel, A.-C., Buchberger, G., and Comanns, P. (2017), "Moisture-Harvesting Reptiles: A Review," in *Functional Surfaces in Biology III: Diversity of the Physical Phenomena* (S. N. Gorb and E. V. Gorb, eds.), pp. 93–106, Springer International, Cham, Switzerland.

Joubert, C. S. W. and Maclean, G. L. (1973), "The Structure of the Water-Holding Feathers of the Namaqua Sandgrouse," *Zoologica Africana* **8**, 141–152.

Ju, J., Bai, H., Zheng, Y., Zhao, T., Fang, R., and Jiang, L. (2012), "A Multi-Structural and Multi-Functional Integrated Fog Collection System in Cactus," *Nat. Commun.* **3**, 1247.

Kidron, G. J. (2005), "Angle and Aspect Dependent Dew and Fog Precipitation in the Negev Desert," *J. Hydrol.* **301**, 66–74.

Kidron, G. J., Herrnstadt, I., Barzilay, E. (2002), "The Role of Dew as a Moisture Source for Sand Microbiotic Crust in the Negev Desert Israel," *J Arid Environ* **52**, 517–533.

Koch, K., Bhushan, B., and Barthlott, W. (2008), "Diversity of Structure, Morphology and Wetting of Plant Surfaces," *Soft Matter* **4**, 1943–1963.

Laity, J. (2008), *Deserts and Desert Environments*, Wiley–Blackwell Publishing, Hoboken, New Jersey.

Lange, O. L., Meyer, A., Ullmann, I., and Zellner, H. (1991), "Microclimate Conditions, Water Content and Photosynthesis of Lichens in the Coastal Fog Zone of the Namib Desert: Measurements in the Fall," *Flora* **185**, 233–266.

Lillywhite, H. B. and Licht, P. (1974), "Movement of Water over Toad Skin: Functional Role of Epidermal Sculpturing," *Copeia* **1974**, 165–171.

Lillywhite, H. B. and Stein, B. R. (1987), "Surface Sculpturing and Water Retention of Elephant Skin," *J. Zool.* **211**, 727–734.

Liu, C., Xue, Y., Chen, Y., and Zheng, Y. (2015), "Effective Directional Self-Gathering of Drops on Spine of Cactus with Splayed Capillary Arrays," *Sci. Rep.* **5**, 17757.

Louw, G. N. (1972), "The Role of Advective Fog in the Water Economy of Certain Namib Desert Animals," *Symp. Zool. Soc. Lond.* **31**, 297–314.

Louw, G. N. and Seely, M. K. (1980), "Exploitation of Fog Water by a Perennial Namib Dune Grass, Stipagrotis Sabulicola," *S. Afr. J. Sci.* **76**, 38–39.

Malek, E., McCurdy, G., and Giles, B. (1999). "Dew Contribution to the Annual Water Balances in Semi-arid Desert Valleys," *J. Arid Environ.* **42**, 71-80.

Malik, F. T., Clement, R. M., Gethin, D. T., Krawszik, W., and Parker, A. R. (2014), "Nature's Moisture Harvesters: A Comparative Review," *Bioinspir. Biomim.* **9**, 031002.

Malik, F. T., Clement, R. M., Gethin, D. T., Beysens, D., Cohen, R. E., Krawszik, W., and Parker, A. R. (2015), "Dew Harvesting Efficiency of Four Species of Cacti," *Bioinspir. Biomim.* **10**, 036005.

Mares, M. A. (Ed.) (1999), *Encyclopedia of Deserts*, University of Oklahoma, Norman, Oklahoma.

Martins, A. F., Bennett, N. C., Clavel, S., Groenewald, H., Hensman, S., Hoby, S., Joris, A., Manger, P. R., and Milinkovitch, M. C. (2018), "Locally-Curved Geometry Generates Bending Cracks in the African Elephant Skin," *Nat. Commun.* **9**, 1–8.

McClanahan, L. and Baldwin, R. (1969), "Rate of Water Uptake through the Integument of the Desert Toad, Bufo Punctatus," *Comp. Biochem. Physiol.* **28**, 381–389.

Meigs, P. (1953), "World Distribution of Arid and Semi-Arid Homoclimates," in *Reviews of research on arid zone hydrology*, pp. 203–209, United Nations Educational, Scientific, and Cultural Organization, Arid Zone Programme, Paris, France.

Meigs, P. (1966), *Geography of Coastal Deserts*, United Nations Educational, Scientific and Cultural Organization (UNESCO), NS.64/III.33/A, Paris, France.

Monteith, J. L. (1963), "Dew: Facts and Fallacies", in *The Water Relations of Plants*, Rutter, A. J., and Whitehead, F. H. (eds.), pp. 37 –56, Wiley, New York.

Mooney, H. A., Gulmon, S. L., and Weisser, P. J. (1977), "Environmental Adaptations of the Atacaman Desert Cactus Copiapoa Haseltoniana," *Flora* **166**, 117–124.

Moran, M. J., Shapiro, H. N., Boettner, D. D., and Bailey, M. B. (2018), *Fundamentals of Engineering Thermodynamics*, Ninth ed., Wiley, New York.

Murphy, J. A. (2012), *Desert Animal Adaptations*, Capstone Press, North Mankata, Minnesota.

Nobel, P. S. (2003), *Environmental Biology of Agaves and Cacti*, first ed., Cambridge University Press, Cambridge, U.K.

Norgaard, T. and Dacke, M. (2010), "Fog-Basking Behaviour and Water Collection Efficiency in Namib Desert Darkling Beetles," *Front. Zool.* **7**, 23.

Norgaard, T., Ebner, M., and Dacke, M. (2012), "Animal or Plant: Which Is the Better Fog Water Collector?," *PLoS One* **7**, e34603.

Ogburn, R. M. and Edwards, E. J. (2009), "Anatomical Variation in Cactaceae and Relatives: Trait Lability and Evolutionary Innovation," *Am. J. Bot.* **96**, 391–408.

Pan, Z., Pitt, W. G., Zhang, Y., Wu, N., Tao, Y., and Truscott, T. T. (2016), "The Upside-down Water Collection System of Syntrichia Caninervis," *Nat. Plants* **2**, 16076.

Parker, A. R. and Lawrence, C. R. (2001), "Water Capture by a Desert Beetle," *Nature* **414**, 33–34.

Peterson, C. C. (1998), "Rain-Harvesting Behavior by a Free-Ranging Desert Horned Lizard (Phrynosoma Platyrhinos)," *Southwest. Nat.* **43**, 391–394.

Pianka, E. R. and Parker, W. S. (1975), "Ecology of Horned Lizards: A Review with Special Reference to Phrynosoma Platyrhinos," *Copeia* **1975**, 141–162.

Pruppacher, H. R. and Klett, J. D. (2010), *Microphysics of Clouds and Precipitation*, second ed., Springer, New York.

Reiter, R., Höftberger, M., Allan Green, T. G., and Türk, R. (2008), "Photosynthesis of Lichens from Lichen-Dominated Communities in the Alpine/Nival Belt of the Alps – II: Laboratory and Field Measurements of CO2 Exchange and Water Relations," *Flora* **203**, 34–46.

Repp, R. A. and Schuett, G. W. (2008), "Western Diamond-Backed Rattlesnakes, Crotalus Atrox (Serpentes: Viperidae), Gain Water by Harvesting and Drinking Rain, Sleet, and Snow," *Southwest. Nat.* **53**, 108–114.

Robinson, D. A. and Hughes, M. D. (1978), "Observations on the Natural History of Peringuey's Adder, Bitis Peringueyi (Boulenger) (Reptilia: Viperidae)," *Annls. Transv. Mus.* **31**, 189–193.

Roth-Nebelsick, A., Ebner, M., Miranda, T., Gottschalk, V., Voigt, D., Gorb, S., Stegmaier, T., Sarsour, J., Linke, M., and Konrad, W. (2012), "Leaf Surface Structures Enable the Endemic Namib Desert Grass Stipagrostis Sabulicola to Irrigate Itself with Fog Water," *J. R. Soc. Interface* **9**, 1965–1974.

Schill, R., Barthlott, W., and Ehler, N. (1973), "Cactus Spines under the Electron Scanning Microscope," *Cact. Succ. J.* **45**, 175–185.

Schwenk, K. and Greene, H. W. (1987), "Water Collection and Drinking in Phrynocephalus Helioscopus: A Possible Condensation Mechanism," *J. Herpetol.* **21**, 134–139.

Seely, M. K. (1979), "Irregular Fog as a Water Source for Desert Dune Beetles," *Oecologia* **42**, 213–227.

Seely, M. K., de Vos, M. P., and Louw, G. N. (1977), "Fog Inhibition, Satellite Fauna and Unusual Leaf Structure in a Namib Desert Dune Plant Trianthema Hereroensis," *S. Afr. J. Sci.* **73**, 169–172.

Shanyengana, E. S., Henschel, J. R., Seely, M. K., and Sanderson, R. D. (2002), "Exploring Fog as a Supplementary Water Source in Namibia," *Atmospheric Res.* **64**, 251–259.

Sherbrooke, W. C. (1990), "Rain-Harvesting in the Lizard, Phrynosoma Cornutum: Behavior and Integumental Morphology," *J. Herpetol.* **24**, 302–308.

Sherbrooke, W. C. (1993), "Rain-Drinking Behaviors of the Australian Thorny Devil (Sauria: Agamidae)," *J. Herpetol.* **27**, 270–275.

Sherbrooke, W. C. (2002), "Phrynosoma Modestum (Round-Tailed Horned Lizard) Rain-Harvest Drinking Behavior," *Herpetol. Rev.* **33**, 310–312.

Sherbrooke, W. C. (2004), "Integumental Water Movement and Rate of Water Ingestion during Rain Harvesting in the Texas Horned Lizard, Phrynosoma Cornutum," *Amphibia-Reptilia* **25**, 29–39.

Shoemaker, V. H., Balding, D., Ruibal, R., and McClanahan, L. L. (1972), "Uricotelism and Low Evaporative Water Loss in a South American Frog," *Science* **175**, 1018–1020.

Sigsbee, R. A. (1969), *Nucleation*, Marcel Dekker, New York.

Silberglied, R. and Aiello, A. (1980), "Camouflage by Integumentary Wetting in Bark Bugs," *Science* **207**, 773–775.

Simmons, R. E., Griffin, M., Griffin, R. E., Marais, E., and Kolberg, H. (1998), "Endemism in Namibia: Patterns, Processes and Predictions," *Biodivers. Conserv.* **7**, 513–530.

Toledo, R. C. and Jared, C. (1993), "Cutaneous Adaptations to Water Balance in Amphibians," *Comp. Biochem. Physiol A* **105**, 593–608.

Tracy, C. R., Laurence, N., and Christian, K. A. (2011), "Condensation onto the Skin as a Means for Water Gain by Tree Frogs in Tropical Australia," *Am. Nat.* **178**, 553–558.

United Nations Environment Programme (1999), *Sourcebook of Alternative Technologies for Freshwater Augmentation in Some Countries in Asia*, Stationery Office Books, Norwich, UK.

Unmack, P. J. (2001), "Fish Persistence and Fluvial Geomorphology in Central Australia," *J. Arid Environ.* **49**, 653–669.

Van Damme, P. (1991), "Plant Ecology of the Namib Desert," *Afrika Focus* **7**, 355–400.

van Rheede van Oudtshoorn, K. and van Rooyen, M. W. (1999), *Dispersal Biology of Desert Plants*, Springer-Verlag, Berlin, Germany.

Veselý, M. and Modrý, D. (2002), "Rain-Harvesting Behavior in Agamid Lizards (Trapelus)," *J. Herpetol.* **36**, 311–314.

Vesilind, P. J. (2003), "Atacama Desert," *National Geographic*, http://ngm.nationalgeographic. com/features/world/south-america/chile/atacama-text.

Von Hase, A., Cowling, R. M., and Ellis, A. G. (2006), "Petal Movement in Cape Wildflowers Protects Pollen from Exposure to Moisture," *Plant Ecol.* **184**, 75–87.

Walker, A. S. (1992), *Desert: Geology and Resources*, U. S. Geological Survey, Denver Colorado.

Withers, P. (1993), "Cutaneous Water Acquisition by the Thorny Devil (Moloch Horridus: Agamidae)," *J. Herpetol.* **27**, 265–270.

Xue, Y., Wang, T., Shi, W., Sun, L., and Zheng, Y. (2014), "Water Collection Abilities of Green Bristlegrass Bristle," *RSC Adv.* **4**, 40837–40840.

Yenmiş, M., Ayaz, D., Sherbrooke, W. C., and Veselý, M. (2016), "A Comparative Behavioural and Structural Study of Rain-Harvesting and Non-Rain-Harvesting Agamid Lizards of Anatolia (Turkey)," *Zoomorphology* **135**, 137–148.

Yetman, D. (2007), *The Great Cacti: Ethnobotany and Biogeography*, The University of Arizona press, Tuscon, Arizona.

Yom-Tov, Y. (1971), "The Biology of Two Desert Snails Trochoidea (Xerocrassa) Seetzeni and Sphincterochila Boissieri," *Isr. J. Ecol. Evol.* **20**, 231–248.

Chapter 3
Selected Water Harvesting Mechanisms—Lessons from Living Nature

Water moves continuously above and below the surface of the Earth. Bodies of water, clouds, evaporation and condensation all are part of the water cycle. Fog is composed of micron-sized water droplets that form when air becomes saturated with water vapor. Fog is a thick cloud that remains suspended in the atmosphere. Dew is the deposit of water droplets that are formed on cold surfaces, with temperature lower than the dew point, by condensation of water vapor in the air. In many plants and animals, living nature uses fog and condensation as a vital source of water, particularly in arid areas that receive little rainfall (Brown and Bhushan 2016; Bhushan 2018, 2019, 2020). Fog and dew always exist when the temperature decreases late at night and in the early morning. There is evidence that over 5000 years ago, hunter–gatherer groups were able to populate arid areas along the southern coast of Peru by utilizing fresh water from fog and condensation, though the collection method is unknown (Beresford-Jones et al. 2015).

Since the earlier attempts by Carlos Espinosa in Chile in 1957, passive net-based fog harvesters have been built in several countries (Klemm et al. 2012). The net is directly exposed to the atmosphere, and the foggy air is pushed through the net by the wind. Fog droplets are deposited on the net which combine to form large droplets and travel by gravity to a collection dish at the bottom. In North America, many organizations such as FogQuest have used 2D nets for fog interception and harvesting that are able to provide a supplemental source of water in arid regions, shown in Fig. 3.1 (Brown and Bhushan 2016; Bhushan 2018 2019). The yearly water collection rates from 3 to 10 L m^{-2} per day have been reported. To increase water collection rates, new designs were developed, such as a tower by the charity Warka Water project, Fig. 3.1 (Brown and Bhushan 2016; Bhushan 2018, 2019, 2020). The nets in these devices are typically made of material and structures that have not been optimized based on bioinspiration for water collection.

In living nature, several plants and animals in arid regions have evolved surface structures and chemistries that enable them to harvest water from fog and/or condensation of water vapor (Bhushan 2018; Gurera and Bhushan 2020). A detailed overview is presented in Chap. 2. Water harvesting mechanisms of selected plants

© Springer Nature Switzerland AG 2020
B. Bhushan, *Bioinspired Water Harvesting, Purification, and Oil-Water Separation*, Springer Series in Materials Science 299, https://doi.org/10.1007/978-3-030-42132-8_3

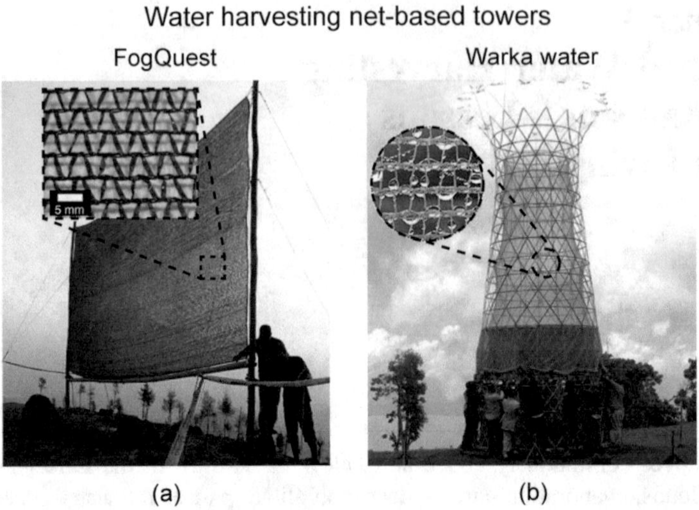

Water harvesting net-based towers

FogQuest   Warka water

(a)   (b)

Fig. 3.1 Net-based water harvesters currently available from FogQuest (photograph by Anne Lummerich) and Warka Water (photograph by Architecture and Vision)

and animals, which are being exploited for commercial applications, are presented in Figs. 3.2 and 3.3 respectively (Brown and Bhushan 2016; Bhushan 2018, 2019, 2020). These include cactus, grass, moss, bushes, Namib desert beetles, lizards, rattlesnakes and spider webs. On their surfaces, droplets from fog or condensation get deposited, and grow before eventually being transported to where it is consumed or stored. A key design feature is that collected water needs to be transported to where it is stored or consumed before it is evaporated. For example, in the case of cactus, grass, moss, or bushes, the collected water is transported towards roots or trunk for storage, and in the case of desert beetles and lizards, collected water is transported towards their mouths for consumption. High droplet speed is needed for water collection purposes to minimize water loss to evaporation.

An overview of water harvesting mechanisms of selected plants and animals follows.

3.1   Cactus

Cactus commonly survive in arid regions. They are mostly succulents, a type of plant with thick, fleshy, and swollen parts that are adapted to store water and minimize water loss (Ogburn and Edwards 2009). Leaves are reduced to spines to reduce water evaporation and protect against feeding animals. Mooney et al. (1977) reported that cacti species use fog as a supplemental source of water. *Copiapoa haseltoniana*, a specie native to the Atacama Desert, utilizes the run off from nightly

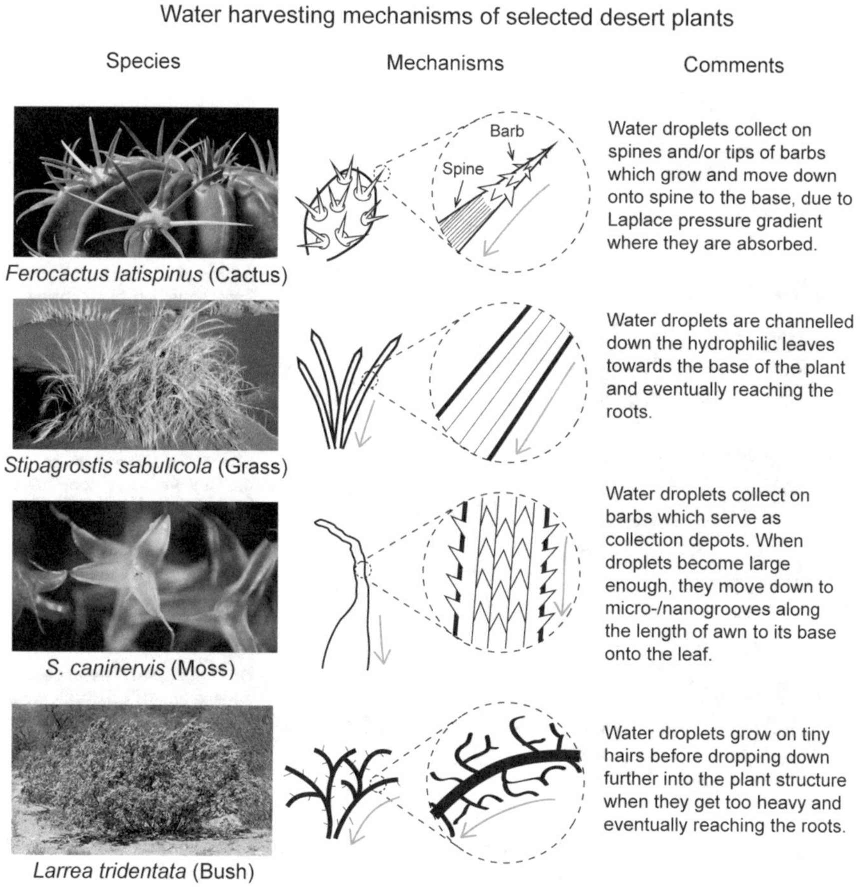

Water harvesting mechanisms of selected desert plants

Species	Mechanisms	Comments
Ferocactus latispinus (Cactus)		Water droplets collect on spines and/or tips of barbs which grow and move down onto spine to the base, due to Laplace pressure gradient where they are absorbed.
Stipagrostis sabulicola (Grass)		Water droplets are channelled down the hydrophilic leaves towards the base of the plant and eventually reaching the roots.
S. caninervis (Moss)		Water droplets collect on barbs which serve as collection depots. When droplets become large enough, they move down to micro-/nanogrooves along the length of awn to its base onto the leaf.
Larrea tridentata (Bush)		Water droplets grow on tiny hairs before dropping down further into the plant structure when they get too heavy and eventually reaching the roots.

Fig. 3.2 Water harvesting mechanisms of selected desert plants which collect water from fog and condensation of water vapor. A combination of surface structure and chemistry results in interception of water from fog and transport to the roots or another area where it can be consumed or stored (adapted from Bhushan 2018, 2019, 2020)

fog events to survive in arid deserts (Vesilind 2003). _Opuntia microdasys_ is a specie of cacti endemic to Mexico, which features small barbs atop conical spines that help in the collection of water (Ju et al. 2012; Bhushan 2018). _Ferocactus latispinus_ is another specie which features conical spines with tiny barbs (Bhushan 2019, 2020). _Gymnocalycium baldianum_, a specie in neighboring Argentina, collects and transports water through microcapillaries (Liu et al. 2015).

Figure 3.2 shows a schematic of spines with microscopic conical barbs on _Ferocactus latispinus_ specie (Bhushan 2018, 2020). Water droplets collect on the conical spine and move towards the base due to curvature gradient providing Laplace pressure gradient. Droplets may also collect on the tips of the small barbs and once they reach critical size, they move onto the conical spine. Once at the base

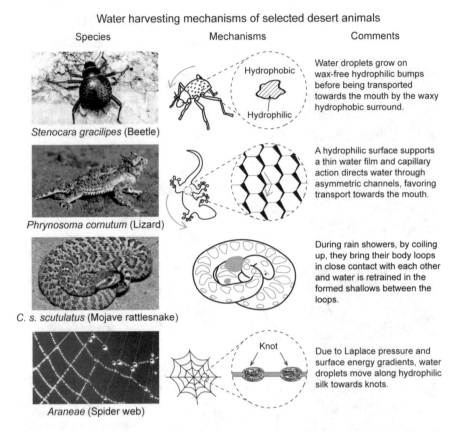

Fig. 3.3 Water harvesting mechanisms of selected desert animals which collect water from fog and condensation of water vapor. A combination of surface structure and chemistry results in interception of water from fog and transport to the mouth or another area where it can be consumed or stored (adapted from Bhushan 2018, 2019, 2020)

of the spine, the plant absorbs the water (Brown and Bhushan 2016; Bhushan 2018). Laplace pressure gradient is large enough that water droplets can defy gravity and climb upward, as shown in Fig. 3.4 (Bhushan 2020). Analysis of Laplace pressure gradient acting on a conical surface with curvature gradient is presented in Appendix 3.A (Lorenceau and Quéré 2004; Bhushan 2018).

In addition to the presence of Laplace pressure gradient on a conical object, the surface slope changes the gravitational forces acting on the surface. Therefore, for a conical object, with tip pointed outward, and inclined with an angle higher than 0° with respect to the horizontal axis, droplet movement is facilitated both by Laplace pressure gradient and gravitational forces.

A number of water collection experiments have been performed on conical surfaces and on triangular geometry on flat surfaces to demonstrate that Laplace pressure gradient assists in droplet transport which increases the water collection rate (Bhushan 2018, 2019, 2020).

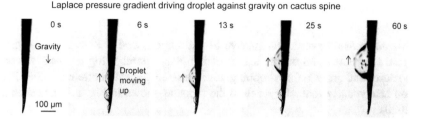

Fig. 3.4 Photographs of water droplets initially collected at the barb tip and/or spines of *Ferocactus latispinus*, climbed up over the cactus spine due to Laplace pressure gradient. Droplets defy gravity (adapted from Bhushan 2020)

3.2   Grass

Other plant species adapted to collect water from fog include *Stipagrostis sabuli-cola*, a grass endemic to the Namib desert (Brown and Bhushan 2016; Bhushan 2018). Water droplets collect on the leaf before coalescing and running down towards the base of the plant (Louw and Seely 1980; Ebner et al. 2011; Roth-Nebelsick et al. 2012). The leaves feature longitudinal ridges, which channel the fluid flow, shown in Fig. 3.2. Another type of grass, *Setaria viridis*, is found to collect water with a similar structure and mechanism which is the same as that in the *Opuntia microdasys* cacti (Xue et al. 2014).

3.3   Desert Moss

Desert moss can survive in extremely arid regions. *Syntrichia caninervis* is a common desert moss. This plant is unique because its leaf surfaces must be wet for photosynthesis to occur, and its root-like structures do not collect water from the soil. The 0.5–2 mm long hair-like structures, referred to as awns or trichomes, exist at the tip of each leaf of the *S. carninervis*, which are used to collect water from fog, dew and rain (Fig. 3.3) (Koch et al. 2008; Pan et al. 2016; Bhushan 2018, 2020).

Awns are hydrophobic and have multiscale structure which facilitate water collection and transport. The awn is covered with nano- and micro-scale grooves and larger scale elongated conical barbs. Barbs serve as collection depots where collected water forms small droplets. When droplets become large enough, they move downward due to Laplace pressure gradient through the micro-/nano-groves present along the length of the awn. Droplets then move towards its base on the leaf surface. The awns provide water collection, droplet formation, and rapid trans-portation to the leaf to keep the moss alive (Bhushan 2020).

3.4   Bushes

Larrea tridentata bush can survive in arid regions and is known to collect water from fog. Fine hairs on the leaves of the plant intercept fog droplets where they coalesce and grow before dropping down into the plant structure when they become too heavy, and eventually reach to the roots, as shown in Fig. 3.2 (Bhushan 2018, 2020).

3.5   Namib Desert Beetles

The Namib desert in southern Africa is one of the most arid regions in the world with average annual rainfall of only on the order of 18 mm, and it is not uncommon to experience consecutive years with no rainfall at all (Shanyengana et al. 2002). *Stenocara gracilipes* and *Onymacris unguicularis* are beetles native to this region. The beetles survive as a result of collection of water from fog. The first observation of fog harvesting in the Namib desert was made by Hamilton and Seely (1976), with beetles emerging during nocturnal fogs and lowering their heads while oriented into the wind. The water was found to trickle down the body of the beetle and into the mouth.

The back of the beetle is comprised of a random array of about 0.5 mm diameter bumps spaced 0.5–1.5 mm apart, shown in Fig. 3.3. The bumps are smooth, while the surrounding area is covered in microstructured wax (Parker and Lawrence 2001). The bumps are hydrophilic while the background wax is hydrophobic. Water from the fog lands on the bumps and droplets begin to grow. The droplet continues to grow (up to about 5 mm) until the weight of the droplet overcomes the capillary force and the droplet detaches and gets transported to the neighboring hydrophobic region and eventually rolls down the tilted beetle's back to its mouth (Bhushan 2018, 2019, 2020). Definition of various wetting states are presented in Appendix 3.B.

It has been shown that superhydrophobic surfaces with superhydrophilic spots increase the fog water collection as compared to the flat surfaces with various homogeneous wettability (Bhushan 2018, 2019, 2020). These experiments verified that an array of hydrophilic bumps on a hydrophobic background is efficient for fog harvesting.

3.6   Lizards

Moloch horridus is a lizard specie native to arid regions in western and southern Australia, shown in Fig. 3.3. Water droplets deposited on the body spread out over the skin before reaching the mouth (Bentley and Blumer 1962). Water movement

Directional spreading of water droplet on lizard skin

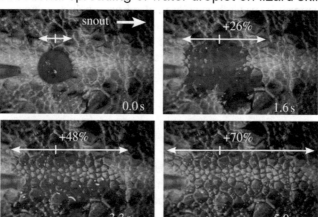

Fig. 3.5 Time-lapse images showing directional spreading of water droplets deposited on lizard skin. Preferential water transport occurs towards the snout (adapted from Comanns et al. 2015)

occurs due to capillary action along open channels in the skin. The lizard adopts a posture where the head is depressed and the hindquarters are elevated which helps guide the water to the mouth, as reported for *Phrynocephalus helioscopus*, a species of lizard native to arid regions in Asia (Schwenk and Greene 1987; Bhushan 2018, 2019, 2020). Similarly, in the case of *Phrynosoma cornutum* or the Texas horned lizard, water placed on the skin flowed preferentially towards the mouth, with capillary forces dominating over gravitational and viscous forces, shown in Fig. 3.5 (Comanns et al. 2015). A network of capillary channels exist with a narrowing of individual channels in the direction of the mouth (longitudinal), shown in Fig. 3.3 (Bhushan 2018). An abrupt widening from one channel to the next, in addition to an interconnecting narrower channel running laterally, would follow. The narrowing of the channel results in favorable water transport in that direction due to the curvature of the liquid–air interface. The lateral interconnecting channels overcome the effects of the abrupt widening and help maintain an advancing liquid front. In the backward direction, liquid flow is stopped as the channel widens and pressure would need to be applied to force the liquid in that direction (Brown and Bhushan 2016; Bhushan 2018, 2019, 2020).

3.7 Rattlesnakes

Many moisture-harvesting reptiles collect water from fog. For example, desert snakes emerge from their dens or any potential roofing during rain showers and coil up in the open. By coiling up, they bring their body loops in close contact with each other and water is retained in the formed shallows between the loops, shown in Fig. 3.3 (Cardwell 2006; Gorb and Gorb 2017; Bhushan 2020; Gurera and Bhushan 2020). Some species also exhibit a dorsoventral flattening of the body, probably to increase exposed surface area. With the snout continuously in contact with the shallows and the head moving contrary to scales' overlap, the collected water is ingested.

3.8 Spider Webs

Spider webs are known to collect water, as evidenced by capturing a dew-glistened web, shown in Fig. 3.3 (Brown and Bhushan 2016; Bhushan 2018). Dry webs undergo moisture-induced structural reorganization. First, the hygroscopic nature of the proteins contained within the silk results in the condensation of water droplets and the swelling of the cylindrical silk thread. This cylinder is then broken up due to Rayleigh instability, where a cylinder of fluid will break up into smaller drops to lower its surface area, resulting in the formation of a "beads on a string" structure with a series of knots periodically spaced along the thread, shown in Fig. 3.6a (Edmonds and Vollrath 1992; Brown and Bhushan 2016; Bhushan 2018). It is believed that this rebuilding of the web structure and subsequent water capture is beneficial for the spider, as not only does it provide a source of drinking water, but it also results in improved capture of prey due to the water-swollen knots possessing enhanced adhesive properties (Edmonds and Vollrath 1992).

Web knots are known to be composed of randomly oriented porous nanofibrils, while the interconnecting joints are composed of stretched porous nanofibrils aligned parallel to the thread, shown in Fig. 3.6b–d (Zheng et al. 2010; Brown and Bhushan 2016; Bhushan 2018). When water condenses on the wet-rebuilt silk fiber, the droplets condensing on the joints are found to move to the knots, shown in Fig. 3.6e–j. A combination of a surface tension gradient and a Laplace pressure gradient is believed to be responsible for the water movement (Zheng et al. 2010). Surface tension gradient occurs due to the knots displaying a rougher surface because of the randomly oriented nanofibrils. The roughness enhances their hydrophilicity, since the droplets are in the Wenzel state of wetting. Laplace pressure gradient occurs due to the larger radius of curvature of the joints compared to that of knots. Droplets, therefore, move on a conical fiber from regions of smaller radius to regions of larger radius (Brown and Bhushan, 2016; Bhushan, 2018).

3.9 Summary

Many plants and animals commonly found in arid regions exhibit the ability to harvest water from fog and condensation of water vapor. Water harvesting mechanisms of selected plants and animals are presented in Figs. 3.2 and 3.3 (Brown and

Water droplet collection on spider webs

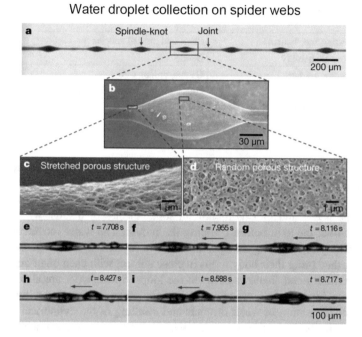

Fig. 3.6 a Image of spider web after water-induced structural changes, comprising thick knots connected by thinner joints due to Rayleigh instability; **b–d** SEM images showing morphology of knot and joint; and **e–j** time-lapse images showing preferential transport of water droplets from joints to knots on a silk fiber (adapted from Edmonds and Vollrath 1992; Zheng et al. 2010; Brown and Bhushan 2016; Bhushan 2018)

Bhushan 2016; Bhushan 2018, 2019, 2020). These examples typically contain a combination of surface structure and chemistry to achieve efficient interception, transport, and collection of water. A key design in consideration is to transport water to where it is stored or consumed before it is evaporated. Species use various methods including heterogeneous wettability, longitudinal grooves and conical or triangular geometry (to drive droplets by Laplace pressure gradient). A summary is presented in Table 3.1 (Gurera and Bhushan 2020).

In the case of cactus, water droplets collect on tips of small conical barbs and/or spines and grow and move down onto spine towards the base due to Laplace pressure gradients. Certain grasses contain grooved surfaces with anisotropic wetting characteristics, which direct water down towards the roots. In the case of desert moss, water droplets collect on the barbs which serve as collection depots. When droplets become large enough, they move down to the micro-/nano-groves along the length of awn. Droplets then move the base onto the leaf surface. Some bushes contain fine hairs on the leaves of the plant which intercept fog droplets where they coalesce, grow, and drop down into the plant structure when they become heavy, eventually reaching to the roots.

Table 3.1 Summary of selected water harvesting mechanisms with potential in commercial applications

Water harvesting mechanisms	Plant and animal species
Heterogeneous wettability	Beetle
Laplace pressure gradient	Cactus, spider web
Grooves	Grass, cactus, moss, elephant, lizard, toad
Water retention in shallows	Rattlesnakes
Coalescence and gravity	Bush

A desert beetle shell comprises a random array of bumps. These bumps are hydrophilic, while the rest of the beetle back is hydrophobic. Therefore, water droplets collect on these bumps and once they are large enough, travel over the body of the beetle and into its mouth. In some species of lizards, water droplets spread out over the skin before reaching the mouth. The water movement occurs due to capillary action along the open channels in the skin. During rain showers, by coiling up, they bring their body loops in close contact with each other and water is retained in the formed shallows between the loops. Spider webs are also known to collect water. First, the hygroscopic nature of the proteins contained within the silk results in condensation of water droplets and swelling of the cylindrical silk fiber. This results in the formation of water-swollen knots with interconnecting joints. Compared to the joints, the knots are rougher with higher hydrophilicity and have a smaller radius, which produce a combination of surface energy and Laplace pressure gradients, respectively. Therefore, further condensation of water droplets on the joints move these to the knots of a conical fiber.

Selected dimensions of cactus, desert grass and desert beetle are presented in Table 3.2 (Gurera and Bhushan 2019).

Table 3.2 Selected dimensions describing three water collecting species (adapted from Gurera and Bhushan 2019)

Water collecting specie	Dimensions and contact angles
Cactus[a]	Spine length ~ 1.5 mm, base diameter ~ 50 μm, tip angle—$10°$, barbed length—top 1/4th of the spine length, groove length—bottom 3/4th of the spine length, groove width ~ 2 μm, groove pitch ~ 20 μm, contact angle $\sim 120°$
Desert grass[b]	Width ~ 2 mm, length <2000 mm, groove width ~ 0.3 mm, groove pitch ~ 0.4 mm, contact angle $\sim 75°$
Desert beetle[c]	Hydrophilic spot diameter—0.2 to 0.5 mm, pitch—0.5 to 1.5 mm on a hydrophobic surface ($\sim 105°$)

[a]Ju et al. (2012)
[b]Roth-Nebelsick et al. (2012)
[c]Parker and Lawrence (2001)

Appendix 3.A: Laplace Pressure Gradient on a Conical Surface

For a spherical droplet sitting on a surface, capillary pressure or Laplace pressure in the liquid p_L is proportional to the surface tension of the liquid in air (γ_{LA}) divided by the local radius, R (Adamson and Gast 1997; Bhushan 2013a, b, 2018),

$$p_L = \frac{\gamma_{LA}}{R} \qquad (3.A.1)$$

The Laplace pressure can be attractive or repulsive depending on whether the surface is hydrophilic or hydrophobic, respectively. The p_L remains constant on a flat surface.

Next, we consider a liquid droplet sitting on a conical object. We consider two adjacent locations A and B with the local radii of the cone, as R_A and R_B, respectively, Fig. 3.A.1 (Bhushan 2018). The surface curvature gradient results in the Laplace pressure difference between the two opposite ends of the droplet along the surface. The Laplace pressure difference is given as,

$$\Delta p_L = \gamma_{LA} \left(\frac{1}{R'_A} - \frac{1}{R'_B} \right) \qquad (3.A.2)$$

Fig. 3.A.1 Schematic of a liquid droplet on a conical object with a cone angle of 2α and local radii R_A and R_B at two locations of A and B, respectively. Shown is the Laplace force (F_L) which helps in driving the droplet towards a larger radius (adapted from Bhushan 2018)

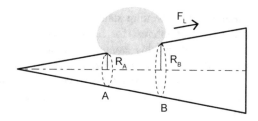

Laplace force acting on a water droplet

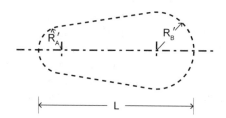

Contact area between water droplet and cone
(Top view)

where R'_A and R'_B are the radii of curvature at the front and rear contact lines of the droplet, respectively. The curvature gradient leading to Laplace pressure difference that acts on the contact area A, produces the Laplace force F_L,

$$F_L = \iint_A \Delta p_L dA \tag{3.A.3}$$

where the contact area is approximately equal to volume of the droplet, $V_{droplet}$ divided by the length L of the droplet,

$$A \sim \frac{\sin \alpha}{(R_B - R_A)} V_{droplet} \tag{3.A.4}$$

where α is the half apex angle of the cone. Combining (3.A.2) to (3.A.5), we get,

$$F_L \sim \gamma_{LA} \left(\frac{1}{R'_A} - \frac{1}{R'_B} \right) \frac{\sin \alpha}{R_B - R_A} V_{droplet} \tag{3.A.5}$$

The Laplace force acting on a conical object drives the droplet from regions of lower radius to larger radius as long as the Laplace force is larger than adhesion force. During this droplet movement, new droplets may be deposited in the path, which coalesce resulting in a large liquid volume and provide the additional movement.

Appendix 3.B: Definition of Various Wetting States

To define various wetting states of surfaces, we start with the definition of a few Greek words of interest for liquids: hydro- = water, oleo- = oil, amphi- = both (water and oil in this context), omni- = all or everything, and liquid- = an unspecified liquid. Within oils, edible oils have a surface tension larger than 30 mN/m and alkanes-based oils have a surface tension 20–30 mN/m. Greek suffixes of interest are: -philic = friendly or attracting, and -phobic = afraid of or repelling.

Figure 3.B.1 shows schematics of various wetting states (Bhushan 2016, 2018). If a liquid wet a surface, it is referred to as a wetting liquid and the value of the static contact angle is $0 \leq \theta \leq 90°$. A surface that is wetted by a wetting liquid is referred to as hydrophilic if that wetting liquid is water, oleophilic if it is oil, amphiphilic if it can be wetted by water and oil, omniphilic if all liquids wet the surface, and liquiphilic if the liquid is unspecified. If the liquid does not wet the surface, it is referred to as a non-wetting liquid, and the value of the contact angle is $90° < \theta \leq 180°$. A surface that repels a liquid is referred to as hydrophobic if it repels water, oleophobic for oil, amphiphobic if it repels both water and oil, omniphobic if it repels all liquids, and liquiphobic if the liquid is unspecified.

Wetting states

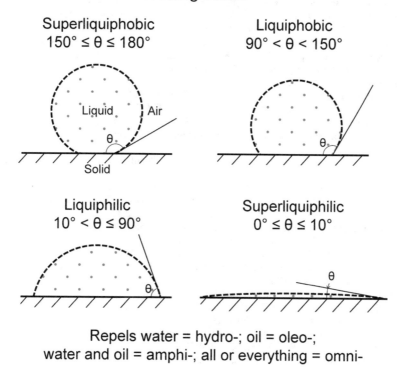

Fig. 3.B.1 Schematics of liquid droplets in contact with superliquiphobic, liquiphobic, liquiphilic, and superliquiphilic solid surfaces (Bhushan 2016)

Surfaces with a contact angle of less than 10° are called superliquiphilic, while surfaces with a contact angle between 150° and 180° are called superliquiphobic. The words superliquiphilic and superliquiphobic were coined by Bhushan (2016).

References

Adamson, A. V. and Gast, A. P. (1997), *Physical Chemistry of Surfaces*, sixth ed., Wiley, New York.

Bentley, P. J. and Blumer, W. F. C. (1962), "Uptake of Water by the Lizard, *Moloch horridus*," *Nature* **194**, 699–700.

Beresford-Jones, D., Pullen, A. G., Whaley, O. Q., Moat, J., Chauca, G., Cadwallader, L., Arce, S., Orellana, A., Alarcón, C., Gorriti, M., Maita, P. K., Sturt, F., Dupeyron, A., Huaman, O., Lane, K. J., and French, C. (2015), "Re-evaluating the Resource Potential of Lomas Fog Oasis Environments for Preceramic Hunter–gatherers under Past ENSO Modes on the South Coast of Peru," *Quat. Sci. Rev.* **129**, 196–215.

Bhushan, B. (2013a), *Introduction to Tribology,* second ed., Wiley, New York.

Bhushan, B. (2013b), *Principles and Applications of Tribology,* second ed., Wiley, New York.

Bhushan, B. (2016), *Biomimetics: Bioinspired Hierarchical-Structured Surfaces for Green Science and Technology,* second ed., Springer International, Cham, Switzerland.

Bhushan, B. (2018), *Biomimetics: Bioinspired Hierarchical-Structured Surfaces for Green Science and Technology,* third ed., Springer International, Cham, Switzerland.

Bhushan, B. (2019), "Bioinspired Water Collection Methods to Supplement Water Supply," *Phil. Trans. R. Soc. A* **377**, 20190119.

Bhushan, B. (2020), "Design of Water Harvesting Towers and Projections for Water Collection from Fog and Condensation," *Phil. Trans. R. Soc. A* **378**, 20190440.

Brown, P. S. and Bhushan, B. (2016), "Bioinspired Materials for Water Supply and Management: Water Collection, Water Purification and Separation of Water from Oil," *Phil. Trans. R. Soc. A* **374**, 20160135.

Cardwell, M. D. (2006), "Rain Harvesting in a Wild Population of *Crotalus s. scutulatus* (Serpentes: Viperidae," *Herpetol. Rev.* **37**, 142–144.

Comanns, P., Buchberger, G., Buchbaum, A., Baumgartner, R., Kogler, A., Bauer, S., and Baumgartner, W. (2015), "Directional, Passive Liquid Transport: the Texas Horned Lizard as a Model for a Biomimetic 'Liquid Diode'," *J. R. Soc. Interface* **12**, 20150415.

Ebner, M., Miranda, T., and Roth-Nebelsick, A. (2011), "Efficient Fog Harvesting by *Stipagrostis sabulicola* (Namib Dune Bushman Grass)," *J. Arid Environ.* **75**, 524–531.

Edmonds, D. T. and Vollrath, F. (1992), "The Contribution of Atmospheric Water Vapour to the Formation and Efficiency of a Spider's Capture Web," *Proc. R. Soc. Lond. B* **248**, 145–148.

Gorb, S. N. and Gorb, E. V. (Eds.) (2017), *Functional Surfaces in Biology III – Diversity of the Physical Phenomena,* Springer International, Cham, Switzerland.

Gurera, D. and Bhushan, B. (2019), "Designing Bioinspired Surfaces for Water Collection from Fog," *Phil. Trans. R. Soc. A* **377**, 20180269.

Gurera, D. and Bhushan, B. (2020), "Passive Water Harvesting by Desert Plants and Animals: Lessons from Nature," *Phil. Trans. R. Soc. A* **378**, 20190444.

Hamilton, W. J. and Seely, M. K. (1976), "Fog Basking by the Namib Desert Beetle, *Onymacris unguicularis,*" *Nature* **262**, 284–285.

Ju, J., Bai, H., Zheng, Y., Zhao, T., Fang, R., and Jiang, L. (2012), "A Multi-structural and Multi-functional Integrated Fog Collection System in Cactus," *Nat. Commun.* **3**, 1247.

Klemm, O., Schemenauer, R. S., Lummerich, A., Cereceda, P., Marzol, V., and Corell, D., et al. (2012), "Fog as a Fresh-Water Resource: Overview and Perspectives," *Ambio* **41**, 221–234.

Koch, K., Bhushan, B., and Barthlott, W. (2008), "Diversity of Structure, Morphology and Wetting of Plant Surfaces," *Soft Matter* **4**, 1943–1963.

Liu, C., Xue, Y., Chen, Y., and Zheng, Y. (2015), "Effective Directional Self-gathering of Drops on Spine of Cactus with Splayed Capillary Arrays," *Sci. Rep.* **5**, 17757 pp 1–8.

Lorenceau, É. and Quéré, D. (2004), "Drops on a Conical Wire," *J. Fluid Mech.* **510**, 29–45.

Louw, G. N. and Seely, M. K. (1980), "Exploitation of Fog Water by a Perennial Namib Dune Grass, *Stipagrotis sabulicola,*" *S. Afr. J. Sci.* **76**, 38–39.

Mooney, H. A., Weisser, P. J., and Gulmon, S. L. (1977), "Environmental Adaptations of the Atacaman Desert Cactus *Oopiapoa haseltoniana,*" *Flora, Bd.* **166**, S. 117–124.

Ogburn, R. M. and Edwards, E. J. (2009), "Anatomical Variation in Cactaceae and Relatives: Trait Lability and Evolutionary Innovation," *Am. J. Bot.* **96**, 391–408.

Pan, Z., Pitt, W. G., Zhang, Y., Wu, N., Tao, Y. and Truscott, T. T. (2016), "The Upside-down Water Collection System of *Syntrichia caninervis,*" *Nat. Plants* **2**, 16076.

Parker, A. R. and Lawrence, C. R. (2001), "Water Capture by a Desert Beetle," *Nature* **414**, 33–34.

Roth-Nebelsick, A., Ebner, M., Miranda, T., Gottschalk, V., Voigt, D., Gorb, S., Stegmaier, T., Sarsour, J., Linke, M., and Konrad, W. (2012), "Leaf Surface Structures Enable the Endemic Namib Desert Grass *Stipagrostis sabulicola* to Irrigate Itself with Fog Water," *J. R. Soc. Interface* **9**, 1965–1974.

Schwenk, K. and Greene, H. W. (1987), "Water Collection and Drinking in *Phrynocephalus helioscopus*: A Possible Condensation Mechanism," *J. Herpetol.* **21**, 134–139.

Shanyengana, E. S., Henschel, J. R., Seely, M. K., and Sanderson, R. D. (2002), "Exploring Fog as a Supplementary Water Source in Namibia," *Atmos. Res.* **64**, 251–259.

Vesilind, P. J. (2003), "Atacama Desert," *National Geographic* (August, 2003) see http://ngm.nationalgeographic.com/features/world/south-america/chile/atacama-text.

Xue, Y., Wang, T., Shi, W., Sun, L., and Zheng, Y. (2014), "Water Collection Abilities of Green Bristlegrass Bristle," *RSC Adv.* **4**, 40837–40840.

Zheng, Y., Bai, H., Huang, Z., Tian, X., Nie, F.-Q., Zhao, Y., Zhai, J., and Jiang, L. (2010), "Directional Water Collection on Wetted Spider Silk," *Nature* **463**, 640–643.

Chapter 4
Bioinspired Flat and Conical Surfaces for Water Harvesting

It has been mentioned in Chap. 3, that beetles use surfaces, with heterogeneous wettability cactus spines and spider silk use conical geometry, and blades of grass use longitudinal grooves to drive water droplets for water transport and storage/use, before they evaporate. Bioinspired surfaces for water harvesting from fog have been inspired by the beetle (Garrod et al. 2007; Bai et al. 2014; Gurera and Bhushan 2019a), grass (Azad et al. 2015; Gurera and Bhushan 2019a, b), and cactus (Ju et al. 2013; Gurera and Bhushan 2019a, b, c, d; Schriner and Bhushan 2019). For water harvesting from condensation of water vapor, cactus- and spider silk-inspired conical surfaces have also been used (Gurera and Bhushan 2020).

It should be noted that cactus- and spider silk-inspired triangular patterns also have been used for water transport (Chap. 5). The crucial thing for a droplet to move inside a triangular pattern is that it should be touching the borders. This is not the case in cones. Laplace pressure gradient always acts on a droplet sitting on a cone. This presents a clear advantage of cones over flat triangular patterns. However, in the case of cones, with the bottom surface being horizontal, gravitational forces will reduce the distance travelled and the speed of a droplet. Gravitational forces will assist if the cone tip is pointed upwards.

Systematic studies of water collection from fog on flat and conical surfaces was carried out by Gurera and Bhushan (2019a, b, c). Heterogeneous wettability, Laplace pressure gradient, gravitational effects, and grooved surfaces were investigated on flat and conical water collectors. The flat surfaces with different wettability and surface roughness as well as heterogeneous wettability were characterized for their water collection abilities. Conical surfaces with different geometry, ungrooved and grooved, and different wettability were characterized for droplet movement and their water collection abilities at different inclination angles. Cones with nonlinear surface profiles, to maximize the effects of Laplace pressure gradient and gravitation effects, were also studied.

A systematic study of water harvesting from condensation of water vapor on conical surfaces was carried out by Gurera and Bhushan (2020). Metallic cones of different tip angle, cone length and surface area were used. Effect of inclination

© Springer Nature Switzerland AG 2020
B. Bhushan, *Bioinspired Water Harvesting, Purification,
and Oil-Water Separation*, Springer Series in Materials Science 299,
https://doi.org/10.1007/978-3-030-42132-8_4

angle and array was also studied. The results were compared with the water collection from fog.

In this chapter, an overview of studies of water harvesting from fog or condensation using flat and conical surfaces carried out by Gurera and Bhushan (2019a, b, c, 2020) is presented. Based on the data, design guidelines for water harvesting towers with heterogeneous wettability and/or consisting of conical arrays are presented (Bhushan 2018, 2019a).

4.1   Experimental Details

This section describes fabrication of various surfaces, and the experimental apparatus used for water collection measurements from fog or condensation (Gurera and Bhushan 2019a, b, c, 2020).

4.1.1   Fabrication of Water Collector Surfaces for Fog

4.1.1.1   Flat Surfaces with Homogeneous and Heterogeneous Wettability

Flat surfaces with homogeneous wettability included superhydrophobicity, hydrophobicity, hydrophilicity, and superhydrophilicity (Bhushan 2017, 2018). Beetle-inspired surfaces consisted of superhydrophilic spots over a superhydrophobic background. A spot size of 0.5 mm diameter and a pitch of 1 mm was chosen, which was inspired from the desert beetle's dimensions, as given in Table 3.1 (Gurera and Bhushan 2019a). To study whether the size of the asperities roughly equal to the size of fog droplets or smaller affects water collection, superhydrophobic surfaces with two surface roughnesses were fabricated. As stated earlier, fog has water droplets of diameter on the order of 10 μm which will interact with the surface based on its surface roughness and surface energy. Therefore, two sizes of nanoparticles were chosen, one comparable to the fog droplet size (10 μm) and another much smaller (7 nm) (Gurera and Bhushan 2019a).

The substrate used was polycarbonate (PC) because it is a tough material and is commonly used in the fabrication of water bottles. PC substrate is hydrophilic. It was made hydrophobic by vapor deposition of fluorosilane (448931, Sigma Aldrich). The 20 mm × 20 mm samples were placed upside down and a droplet of fluorosilane was placed 1 cm below in an enclosure, and was left for 30 min (Brown and Bhushan 2015).

A superhydrophobic surface was fabricated by spray coating a mixture of hydrophobic silica particles and methylphenyl silicone (SR355S, Momentive

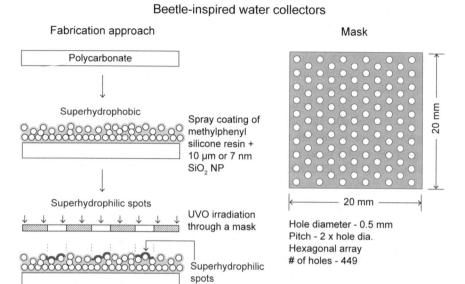

Fig. 4.1 Fabrication of surfaces with heterogeneous wettability for beetle-inspired water collectors (adapted from Gurera and Bhushan 2019a)

Performance Materials) binder on the uncoated PC (Bhushan and Martin 2018; Bhushan 2018, 2019b) (Fig. 4.1). Two different sizes of particles were used— 10 µm (Aerosil VM2270) and 7 nm (Aerosil RX300). The coating mixture was prepared by mixing 375 mg of the particles and 150 mg of the binder in 30 mL of solvent in an ultrasonifier (Branson Sonifier 450A, Emerson Electric Co., St. Louis, Missouri) for 30 min. The solvent used was 40% tetrahydrofuran (THF, Fisher Scientific) and 60% isopropyl alcohol (IPA, Fisher Scientific).

Superhydrophilicity was introduced by treating the PC surface with ultraviolet-ozone (UVO) light. The UVO lamp used was a U-shaped lamp (18.4 W, Model G18T5VH-U, Atlantic Ultraviolet Co.), and the samples were kept directly underneath the light source for 60 min.

To fabricate a beetle-inspired surface, superhydrophilic spots were introduced on the superhydrophobic surface by irradiating the spray coated surface using UVO light through a mask. Schematic of the mask is also shown in Fig. 4.1.

Wettability of surfaces was characterized by measuring static contact angles (CA). The contact angles were measured using an automated goniometer (Model 290, Ramé-Hart Instrument Co.). Five µL distilled water droplets were deposited on to the surface using microsyringe, and the static contact angles were measured by taking static profile images of the water droplets. The profile images were analysed using the DROPimage software.

4.1.1.2   Conical Surfaces With and Without Grooves and Homogeneous and Heterogeneous Wettability

Cactus- and spider silk-inspired conical surfaces with and without grooves (grooves inspired by grass), and with homogeneous or heterogeneous wettability (heterogeneity inspired by beetles), were fabricated using additive manufacturing or 3D printing (Objet30 Prime, Stratasys) that allows flexibility in designing and scalability (Gurera and Bhushan 2019a, b). The machine uses the PolyJet 3D printing technology which is similar to inkjet printing, and jets layers of curable liquid photopolymer onto a build tray. It has an accuracy of 0.1 mm. The material used was acrylic polymer, RGD720. Surface wettability was modified by surface treatment and/or coating deposition.

Cylinder Versus Cone
Conical and cylindrical surfaces were fabricated to study the role of conical geometry. Schematics of cylindrical and conical geometries are presented in Fig. 4.2. The designs were based on the dimensions of natural species, summarized in Table 3.1. A base diameter of the cylinder of 3 mm was chosen in an effort to mimic the grass's diameter. The length was chosen to be 35 mm. To keep the surface area and the length constant, the base diameter of the cone, 6 mm, was doubled to that of cylinder, and the tip angle was kept at 10°, as seen in cactus spines (Gurera and Bhushan 2019a).

Fig. 4.2 Schematics of a 3D printed cylinder and a cactus- and spider silk-inspired cone with equal surface area (adapted from Gurera and Bhushan 2019a)

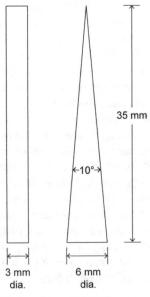

Cylinder and cactus- and spider silk-inspired cone

Material - hydrophilic acrylic polymer, water CA = 61°±2°

35 mm

←10°→

3 mm dia.   6 mm dia.

Surface area = 330 mm^2

Single Cone

(a) Geometry. To study the effect of conical geometry, two tip angles were chosen: 10° and 45°. Cacti conical spines have a tip angle of about 10°, as mentioned in Table 3.1. For comparison, a larger tip angle of 45° was chosen (Gurera and Bhushan 2019b). When considering the collection of water from fog, a comparison between cones of different tip angles and the same surface area is needed because fog intercepts the entire surface and droplets are formed across the whole surface. Therefore, having a similar number of water nucleation sites presents a fair comparison. It is known that a droplet moves from the tip to the base under a fog flow. Therefore, comparing tip angles with same lengths, presents another fair comparison. This results in the traveling distance for droplets to be the same on each cone.

Three cones with two tip angles were fabricated—one pair with the same surface area (330 mm^2) and another pair with the same length (15 mm), as shown in Fig. 4.3a (Gurera and Bhushan 2019b). The surface area, A, is given by $\pi l^2 (\tan(\theta/2)/\cos(\theta/2))$ where l is the cone length, and θ is the tip angle.

(b) Grooves. A representative cone of 45° tip angle and 15 mm length, was chosen to investigate the effect of grooves inspired by desert grass and cactus, as shown in Fig. 4.3b (Gurera and Bhushan 2019b). The number of grooves selected were 8; they were spaced equidistantly, and ran up to 3/4th of the cone's length, as inspired from cactus. The thin grooves found on a cactus spine could not be fabricated using the 3D printer. Therefore, grooves' cross sectional dimension, 0.4 mm by 0.4 mm, inspired from grass (Table 3.1), was used.

(c) Heterogeneity. A representative cone of 45° tip angle and 15 mm length, was chosen to investigate the effect of heterogeneity, as shown in Fig. 4.3b (Gurera and Bhushan 2019b). Heterogeneous wettability in grass and cactus does not exist in nature. It was incorporated with inspiration coming from the desert beetle. Heterogeneous wettability in the conical surfaces included a hydrophobic tip with a superhydrophilic base. A hydrophobic tip was selected as it can collect more water, and a superhydrophilic base because it can transport the collected water quickly to the base. Superhydrophobic tips are not recommended because droplets will not stick to the surface and will fly away instead because of the low tilt angle.

For fabrication, the entire surface of the cones were made superhydrophilic using a UVO lamp. The bottom 3/4th was then covered with Teflon tape and the entire object was subjected to fluorination; afterwards the tape was removed. The fluorination was achieved via vapor deposition of fluorosilane (448931, Sigma Aldrich). The samples were placed upside down and a droplet of fluorosilane was placed 1 cm below the tip in an enclosure, and was left for 30 min (Brown and Bhushan 2015).

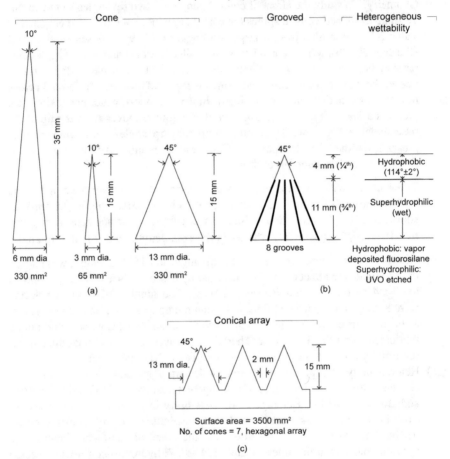

Fig. 4.3 Schematics of 3D printed cactus- and spider silk-, grass-, and beetle-inspired conical water collectors. **a** Single cones with 10° or 45° tip angles having either same length (15 mm) or same surface area (330 mm²), **b** grooved cone with 8 grooves running up to 3/4th of the length starting from the base, and a cone with heterogeneous wettability with bottom 3/4th of the cone superhydrophilic and top 1/4th hydrophobic, and **c** a conical array with 7 cones with 45° tip angle (adapted from Gurera and Bhushan 2019b)

Conical Array

To design arrays, representative cones with 45° tip angle and 15 mm length, were chosen. Seven cones were selected in the array, with end-to-end spacing between two adjacent cones of about 2 mm, as shown in Fig. 4.3c.

Nonlinear Cone

To study the effect of surface profiles of cones, a nonlinear cone was fabricated, shown in Fig. 4.4 (Gurera and Bhushan 2019c). The dimensions of the cone were

Fig. 4.4 Schematics of linear and nonlinear conical water collectors. A nonlinear cone with same length and same base diameter of 45° tip angle cone, but a smaller tip angle of 10° was designed with a concave shape. Two linear cones of the same length and 10° and 45° tip angles were also fabricated for comparison (adapted from Gurera and Bhushan 2019c)

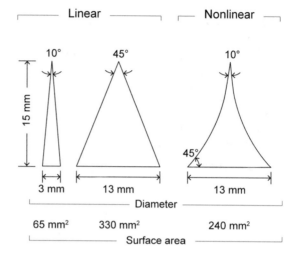

based on two linear cones, with tip angles of 10° and 45°, and length of 15 mm, which are also shown (Gurera and Bhushan 2019b). To design the nonlinear cone, the length of 15 mm and the base diameter of 13 mm of the linear cone of 45° tip angle, but a lower tip angle of 10° were selected. A cubic polynomial was fitted with selected slopes at the tip and at the base.

4.1.2   Fabrication of Water Collector Surfaces for Condensation

Thermally conducting cones are needed for condensation studies. Aluminium single cones and arrays were fabricated for condensation studies (Gurera and Bhushan, 2020).

4.1.2.1   Single Cone

To study the effect of tip geometry, two tip angles were chosen: 10° and 45°. As mentioned earlier, cacti conical spines have a tip angle of 10°. For comparison, a larger tip angle of 45° was also chosen (Gurera and Bhushan 2020).

   When considering the collection of water from condensation, a comparison between cones of same surface area or same length is needed. Condensation is a process in which vapors change into liquid water state and they do that by forming water nucleation site across the surface. Therefore, having a similar number of water nucleation sites presents a fair comparison. It is known that droplets travel from the tip to the base. Therefore, another fair comparison is presented by comparing different tip angles with same cone lengths. This results in the traveling distance for droplets to be the same on each cone (Gurera and Bhushan 2020).

Fig. 4.5 Schematics of cactus- and spider silk-inspired conical water collectors, made of 6061 aluminum. **a** Single cones with 10° and 45° tip angles having either same length (15 mm) or same surface area (330 mm²) and **b** a conical array with 7 cones and 45° tip angle (adapted from Gurera and Bhushan 2020)

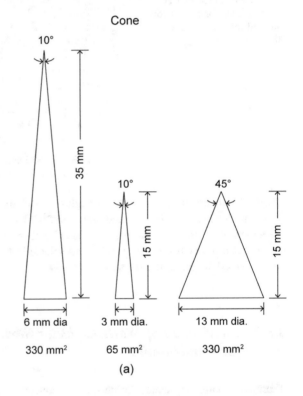

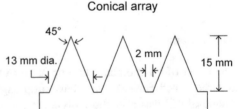

Three cones with two tip angles, were fabricated—a pair with the same surface area (330 mm^2) and another pair with the same cone length (15 mm), as shown in Fig. 4.5a.

Single cones of 6061 aluminum were machined using a CNC lathe (TL-2, Haas, Oxnard, California), using a high-speed steel tool bit. The machined surfaces were rough. When scaling up for mass fabrication CNC may not be a preferable fabrication method. Additive manufacturing (3D printing) is a preferable method for scale up, and they provide smooth surfaces. Therefore, to produce the aluminum cones with surface roughness similar to that of 3D printing, the cones were hand-polished using a sand paper (Part #: 00375337, Crocus fine grade, MSC Industrial, Melville, New York) (Gurera and Bhushan 2020). Surface roughness affects water collection rate (Gurera and Bhushan 2020).

Figure 4.6 (left) presents surface roughness plot of polished aluminum. The surface roughness was characterized using two parameters—root mean square (σ) and peak-to-valley distance (P-V) (Bhushan 2013). The characterization tool was an optical profiler (ZeGage™ 3D Optical Surface Profiler, Zygo, Connecticut). The roughness data for 3D printed acrylic polymer cone is also presented for comparison, Fig. 4.6 (right) (Gurera and Bhushan 2020). Surface roughness of polished aluminum cone is of the same order as that of a 3D printed acrylic cone.

4.1.2.2 Array

A representative cone of 45° tip angle and about 15 mm length was chosen to investigate the effect of array, as shown in Fig. 4.5b. Seven cones were chosen in the array, with end-to-end spacing of about 2 mm, which resulted in a hexagonal array (Gurera and Bhushan 2020).

Surface roughness plots of cones

Machined aluminum (polished)
$\sigma = 0.47$ µm, P-V = 7.7 µm

3D printed acrylic polymer
$\sigma = 1.31$ µm, P-V = 13.1 µm

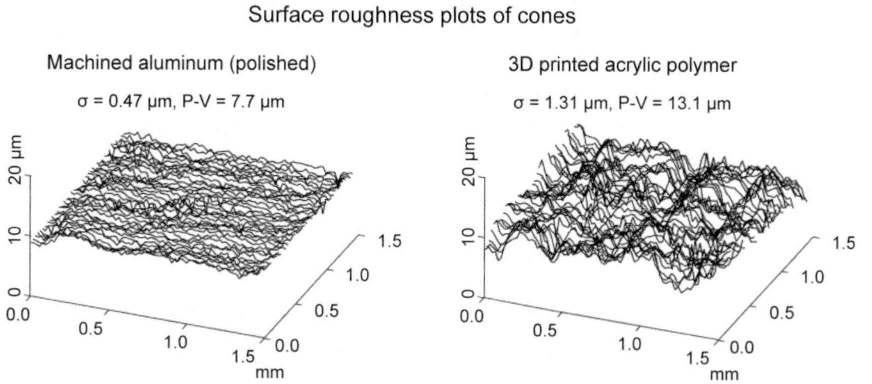

Fig. 4.6 Surface roughness plots of polished aluminum (left) and a 3D-printed acrylic polymer (right). The surface roughness was characterized using two parameters—root mean square (σ) and peak-to-valley (P-V) distance (adapted from Gurera and Bhushan 2020)

For the array, individual cones were manufactured in the same way and they were tack-welded onto a flat plate with predefined holes. The flat plate was also CNC-machined and was made of the same material as of the cones (Gurera and Bhushan 2020).

4.1.3   Experimental Apparatuses for Water Collection

4.1.3.1   Single Droplet Experiments

Single droplet experiments were carried out to understand droplet transport mechanism on cones. The effect of the tip angle of cones and their orientation on the movement of droplets of known volume was studied. Two tip angles were chosen: $10°$ and $45°$. The cones were placed with either sideline or centerline in the horizontal orientation. Sideline in the horizontal orientation was used to study the role of Laplace pressure gradient on the movement of the droplets. Whereas, with centerline in the horizontal orientation, gravitational force also contributed to the movement. Using a pipette, a droplet was placed at the tip of the cone and its motion was photographed (Gurera and Bhushan 2019b; Schriner and Bhushan 2019).

Droplets were fed in increments of 5 µL volume (about 2 mm diameter) which is small enough volume for them to stick to the surface, rather than fall off when ejected from the pipette. Droplets were fed to a total volume of 40 µL, as beyond that volume, droplets fall.

Dependent upon the size of the droplet, it moved for some distance and stopped. After the first droplet had stopped, the pipette was pointed at the current location of the droplet, not the tip of the cone. The increments were added until the droplets detached and fell off the cone surface. Droplets fall because at higher volumes, gravitational forces dominate the capillary forces. For the cones used in this study, droplets detached at deposition of total water volume of about 40 µL.

The distance traveled as a function of droplet volume on the two cones was recorded. The distance was measured from the tip of the cone to the center of the droplet. Experiments were repeated three times for each cone. The average and standard deviation of the distance was reported at every 10 µL increment (Gurera and Bhushan 2019b).

4.1.3.2   Water Collection from Fog

Figure 4.7a shows a schematic of the water collection apparatus used to collect water from fog (Gurera and Bhushan 2019a, b). The conical samples were mounted to a base. The sample base was inclined at either $0°$ or $45°$ from vertical axis (θ) A commercial humidifier (EE-3186, Crane, Itasca, Illinois) was used to produce a stream of fog onto a surface. The falling water droplets were collected in a container and were measured using an analytical balance (B044038, Denver Investment

Fig. 4.7 a Schematic of
experimental apparatus for
water collection from fog.
A commercial humidifier was
used to throw a stream of fog
on a cone sample. The sample
base was inclined at either 0°
or 45° from vertical axis (θ).
Inclination angle of 0° means
that the cone axis is parallel to
the fog flow (horizontal). The
collected water was measured
by an analytical balance
underneath. **b** A
representative water
collection (mg) versus time
(h) plot for hydrophilic flat
surface, and hydrophilic cone
surface of 10° tip angle and
35 mm length, both at 45°
inclination (adapted from
Gurera and Bhushan 2019a,
b)

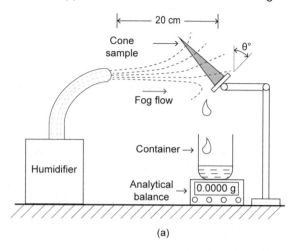

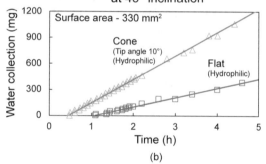

Company, Bohemia, New York) underneath. The minimum mass the balance could
measure was 1 mg. The humidifier which emits fog at about 10 cm/s was kept
about 20 cm away from the surface. The flow speed was calculated by measuring
the volume of water lost over time, and by knowing the diameter of the pipe
through which the fog was blown out.

When the fog is turned on, droplets form all across the intercepted surface area.
There are two forces which drive these droplets to the base—gravity and force due
to the Laplace pressure gradient. To characterize the effect of both forces individually, two inclination angles were chosen—0° and 45°; here 0° means the cone
axis is parallel to the fog flow (horizontal). At 0° inclination angle, the only driving
force is the Laplace pressure gradient. At 45° inclination angle, the force due to
both gravity and the Laplace pressure gradient act on the droplets. In the experiments, mass range of each collected drop was about 30–60 mg.

A representative data of amount of water collection as a function of time for two
surfaces is presented in Fig. 4.7b. The two surfaces were cones with a 10° tip angle

and a flat surface. Both were hydrophilic, had the same surface area of about 330 mm^2, and were inclined at a 45° angle.

4.1.3.3 Water Collection from Condensation

Figure 4.8 shows a schematic of the water collection apparatus used to collect water from condensation (Gurera and Bhushan 2020). The bioinspired cone samples were thermally glued to an aluminum block, with a thermoelectric Peltier cooler (GeekTeches) underneath. The Peltier cooler was used to cool the sample to about 5 ± 1 °C, which is below the dew point. Thermal glue was used to secure the positions (HY910, Halnziye, Shenzhen, Guangdong, China; thermal conductivity: about 1 W/m K). A superhydrophobic plate was used to avoid contact of falling droplets with the aluminum block or the Peltier cooler and for falling droplets to travel to the container. The sample base was inclined at either 0° or 45° from vertical axis (θ) (Gurera and Bhushan 2020).

Every falling water droplet was collected in a container and was weighed using an analytical balance (B044038, Denver Instrument Company, Bohemia, New York). The minimum mass it can measure is 1 mg. In the experiments, the lowest mass of collected water measured was about 10 mg. The typical mass range of each collected drop was about 50–70 mg.

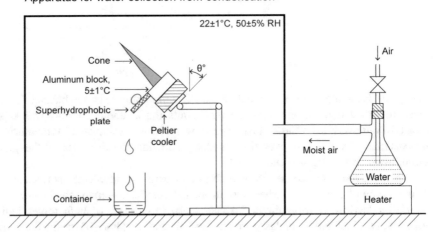

Fig. 4.8 Schematic of experimental apparatus for water collection from condensation. An aluminum cone was thermally glued to an aluminum block with a Peltier cooler underneath. The sample base was inclined at either 0° or 45° from vertical axis (θ). The collected water was measured by an analytical balance underneath. The ambient temperature and humidity was maintained at 22 ± 1 °C and 50 ± 5%, respectively (adapted from Gurera and Bhushan 2020)

The samples were housed in an acrylic chamber (1 m × 0.5 m × 0.8 m). The air temperature in the chamber was 22 ± 1 °C. Relative humidity (RH) was controlled by injection of humid air. The humid air was produced by an air stream that passed through a tank of hot water. By changing the temperature of the hot water and the flow rate of the air stream, RH in the chamber was maintained at about 50%. A digital USB microscope (5 MP, Koolertron, China) was used to capture the condensation process.

4.1.3.4 Water Collection Measurements

Water collected by a cone was measured as a function of time. A straight line was fitted through the set of points and the slope of the lines is referred to as the water collection rate (mg h^{-1}). This parameter was used throughout the study to estimate the water collection ability of different surfaces. A minimum of five droplets were allowed to fall before a straight line was fitted and its slope was calculated. The parameter is believed to be a more accurate representation of the water collection rate, as compared to the other studies which calculated the rate from the beginning of the measurements. The slopes were calculated for a minimum of three different trials. The average and the standard deviation, calculated from a minimum of fifteen data points, were reported (Gurera and Bhushan 2019a, b, 2020).

There are two other parameters by which the water collection data can be characterized. First is the frequency of the droplets falling (droplets h^{-1}), which is the sum of the inverse of the wait-time for every droplet except the first droplet. This parameter gives an idea of how fast the surface is collecting water. The second parameter is average mass of the collected droplet (mg). It is the average of every droplet dropped in the beaker. This could be measured since mass of every droplet was measured individually. Ideally, high frequency and high average droplet mass are desired.

4.2 Results and Discussion—Fog Water Collection Studies

Water collection data from fog for flat, cylindrical, and conical surfaces are presented in this section (Gurera and Bhushan 2019a, b). First, water collection data by flat surfaces of various wettability and beetle-inspired surfaces at two inclination angles and surface roughness (by changing size of nanoparticles) is presented. To study the role of conical geometry, this is followed by a cylinder versus cone with same surface area and two inclination angles. Next data on single droplet transport experiments on two cones with different orientations are presented, followed by water collection data from fog on cones with various geometries and wettability at two inclination angles, and conical arrays. Finally, design guidelines for water collection tower are presented.

4.2.1 Flat Surfaces with Various Wettability and Beetle-Inspired Surfaces at 45° and 0° Inclination Angles

To study the effect of various wettability and heterogeneity on flat surfaces, water collection rate per unit area was measured at a 45° inclination angle. The data are shown in Fig. 4.9a. Superhydrophobic and beetle-inspired surfaces were prepared using 10 μm and 7 nm particles. The size of the particles affects the surface

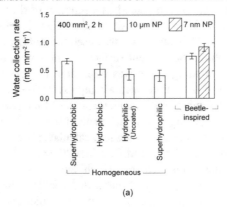

Fig. 4.9 a Water collection rates per unit surface area for flat surfaces with various wettability and beetle-inspired surfaces, at 45° inclination. The wettability includes superhydrophobic, hydrophobic, hydrophilic (uncoated), and superhydrophilic surfaces, and a beetle-inspired surface (which includes 0.5 mm diameter circular superhydrophilic spots surrounded by a superhydrophobic surface). The superhydrophobic and beetle-inspired surfaces were created using two different sizes of NP—10 μm and 7 nm. **b** Optical images showing differences in droplet formation on various surfaces after about 1 h (adapted from Gurera and Bhushan 2019a)

morphology and droplet interaction. Figure 4.9b shows differences in size and shape of droplets on various surfaces after 1 h (Gurera and Bhushan 2019a).

The data shows that the water collection rate increases with higher repellency (higher CA) on the flat surface. This is attributed to the fact that the higher the repellency, the more spherical the droplet shape will be, which results in a lower contact area between the droplet and the surface. Therefore, less heat will be transferred to the droplet and less evaporation will be observed. As shown in the optical images, large spherical droplets are formed on the superhydrophobic surface which roll down. In the case of hydrophilic and superhydrophilic surfaces, the droplets spread across the surface leading to higher evaporation (Gurera and Bhushan 2019a).

A beetle-inspired surface was found to collect more water than flat surfaces with various homogeneous wettability. In beetle-inspired surface, because of the heterogeneity, droplets can slide/roll at a faster rate, and maintain a spherical droplet shape. The faster rate also leads to less evaporation (Gurera and Bhushan 2019a).

Next, the effect of smaller particles was investigated. A coating with smaller particles results in a surface with lower pitch of asperities as compared to the larger particles, which affects the number of nucleation sites. A coating with smaller particles is believed to increase the water collection rate because of more nucleation sites leading to a larger number of smaller droplets coalescing faster. There was an increase in the water collection on the beetle-inspired surface with 7 nm particles. However, the superhydrophobic surface with 7 nm particles resulted in no water collection. This is because of the fact that smaller particles nucleate smaller droplets, as shown in the optical images, which could lead to losing droplets to the wind (Gurera and Bhushan 2019a).

Gurera and Bhushan (2019a) also performed experiments on various surfaces at 0° inclination angle. The data is summarized at two inclination angles in Table 4.1. Water collection rates were lower at 0° inclination angle because of the differences in fog interception. At 0° inclination angle, fog forms a whirl and some droplets get lost to the wind. Whereas at 45° inclination angle, fog intercepts and slides tangentially leading to deposition of larger number of droplets. In the case of the superhydrophobic surface using 7 nm particles, at 0° inclination angle, water collection rate was large, similar to that for a surface using 10 μm particles. It was because the droplets were not lost to the wind at 0° inclination, instead they fell in the beaker (Gurera and Bhushan 2019a).

To summarize, it was found that an increase in water repellency on flat surfaces resulted in higher water collection rates. Water collection rates at 45° inclination angle (with respect to the vertical axis) was larger than at 0° inclination angle. Surfaces with heterogeneous wettability had higher water collection rate than surfaces with homogeneous wettability.

Table 4.1 Summary of water collection rates by flat surfaces with various wettability and beetle-inspired surfaces at two inclination angles, and some fabricated with two nanoparticle (NP) sizes (adapted from Gurera and Bhushan 2019a)

Wettability		Water collection rate (mg mm^{-2} h^{-1})			
		Inclination angle			
		45°		0°	
		10 μm NP	7 nm NP	10 μm NP	7 nm NP
Homogeneous	Superhydrophobic	0.7 ± 0.1	Negligible	0.5 ± 0.1	0.5 ± 0.1
	Hydrophobic	0.5 ± 0.1	–	0.4 ± 0.1	–
	Hydrophilic (Uncoated)	0.4 ± 0.1	–	0.3 ± 0.1	–
	Superhydrophilic	0.4 ± 0.1	–	0.3 ± 0.1	–
Beetle-inspired	0.5 mm spot diameter	0.8 ± 0.1	0.9 ± 0.1	0.7 ± 0.1	0.8 ± 0.1

4.2.2 Cylinder Versus Cone at 0° Inclination Angle

To study the role of conical geometry, water collection rates on a cylinder and a cone were measured. Figure 4.10a shows the water collection data for single cylindrical and conical surfaces, with same surface area and length, at 0° inclination angle (Gurera and Bhushan 2019a). The 0° angle was chosen to minimize the effect of gravity on the movement of the droplets along the length. The left half of the bar chart shows water collection for the complete object. However, in nature, the collected water that matters is that which reaches the base, therefore, the right half of the bar chart presents the water collection rate per unit area at base half-length. The complete water collection rate per unit area by cylinder and cone was found to be comparable. However, it was negligible at the base half-length for cylinder. Figure 4.10b shows comparison of mass and frequency of droplet from a cylinder and a cone in complete collection. Cones collect heavier droplets with a lower frequency.

Figure 4.10c shows optical images of droplets growing and moving on a cylinder and a cone at three time steps (Gurera and Bhushan 2019a). On the cylinder, water droplets nucleate at the tip, grow, attain a critical size, then fall. Droplets mostly nucleate at the tip because they intercept the fog flow perpendicularly. On the rest of the curved surface, little to no droplets were observed since the surface is parallel to the fog flow. The smaller droplets became visible much later; at about the 4 h mark. Since the droplets are small and sit on the surface for a long time, a significant amount of their volume is lost to evaporation. Therefore, the droplets nucleating at the cylinder's tip are mostly responsible for the water collection rate per unit surface area (Gurera and Bhushan 2019a).

The droplets growing at the cylinder's tip do not move along the length, since there is no force acting in that direction. However, this is not the case in cones.

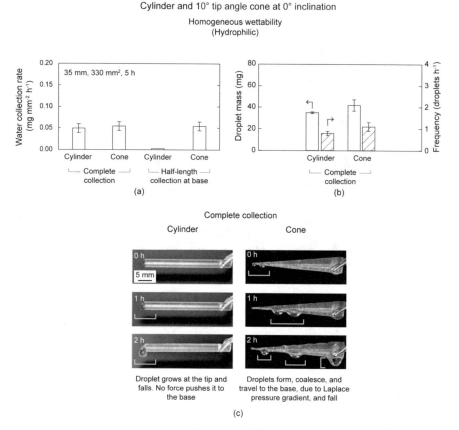

Fig. 4.10 a Water collection rates for a cylinder and a cone at 0° inclination. The left half of the bar chart shows the data when water was collected under the entire surface (complete collection). The right half shows the data when the water was collected under the half-length near the base (as in nature), **b** mass and frequency of falling droplets from a cylinder and a cone in complete collection, and **c** the optical images showing movement of the water droplets on a cylinder and a cone in complete collection (adapted from Gurera and Bhushan 2019a)

Droplets nucleate at the tip, grow and move along its length. For a droplet resting on a conical surface, the changing radius results in a pressure difference inside the droplet (Appendix 3.A). Laplace pressure gradient causes droplets to move on a cone from regions of low radius to regions of higher radius.

Irrespective of the motion of the droplet, both a cylinder and a cone have similar water collection rate per unit area. As mentioned earlier, in nature, the water collection that matters is at the base. A cylinder does not achieve any water collection at the base, whereas a cone achieves the same water collection at the base that was observed in complete collection. As a result, cones have an advantage over cylinders in water collection at the base.

Droplets movement on cone and cylinder

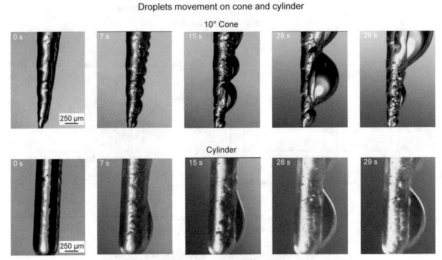

On cone, droplets coalesce and move upward due to Laplace pressure gradient.
On cylinder, droplets grow but do not move.

Fig. 4.11 Video stills of water droplets on a cone with 10° tip angle, pointed downward, and a cylinder, both inclined in the vertical direction. Water droplets grew larger but did not move on a cylinder; however, droplets grew, coalesced and climbed upward on a cone due to Laplace pressure gradient, defying gravity

To illustrate that Laplace pressure gradient defies gravity, Fig. 4.11 shows video stills of droplets on a cone with 10° tip angle pointed downward and a cylinder. It can be seen that droplets grew larger but did not move on a cylinder. However, droplets grew, coalesced and climbed upward on a cone, defying gravity.

4.2.3 Cones at 45° Inclination Angle and Comparison with Flat Surfaces

To study the effect of inclination angle, water collection rates on a cone at 45° inclination angle were measured at the base. Data are shown in Fig. 4.12 (Gurera and Bhushan 2019a). The water collection rate at 45° inclination is about an order of magnitude higher that at 0° inclination (Fig. 4.10). At 45° inclination angle, the entire surfaces intercept the fog flow, not just its tip. Droplets nucleate, grow, slide along the length, and drop at the base. Droplets fall at the base because there is gravity (g sinθ), in addition to Laplace pressure gradient, aiding the movement of the droplets towards the base. It was reported by Gurera and Bhushan (2019a) that at 45° inclination angle, the frequency of the droplets falling was about 6–8 times higher as compared to the 0° inclination. However, aid by gravity did not increase

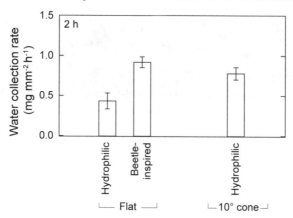

Fig. 4.12 Summary of water collection rates for flat hydrophilic and beetle-inspired surfaces and a hydrophilic cone at 45° inclination (adapted from Gurera and Bhushan 2019a)

the mass of the droplets falling. To summarize, higher inclination increased the water collection rate.

The cone data is also compared with the data for flat and beetle-inspired surfaces in Fig. 4.12. Cones and beetle-inspired surfaces collect about twice as much water as that for a hydrophilic flat surface. Water collection for cones and beetle-inspired surfaces is comparable. Water is transported faster on cones and beetle-inspired surfaces than on the flat surfaces which also reduces evaporation. Finally, in the case of cones, a surface covered with cones (conical array) would increase the surface area to increase collection by several fold, as much as factor of 10 or even more.

4.2.4   Single Droplet Experiments on Cones

A droplet with a volume of 5 µL was deposited at the tip of a 10° tip angle cone. The droplet moved some distance and then came to a stop. If a droplet sits for some time, it will evaporate in a finite time as shown in Fig. 4.13 (Schriner and Bhushan 2019). The droplet with a volume of 5 µL completely evaporated after about 45 min. It is expected that the evaporation rate in real-world scenarios can be higher at high ambient temperatures, low humidity and the presence of wind. As such, it is critical to move droplets from the tip to the base as quickly as possible so that water lost to evaporation is minimal.

To understand droplet transport mechanism in cones, single droplet experiments were conducted on cones with two tip angles of 10° and 45° with sideline or

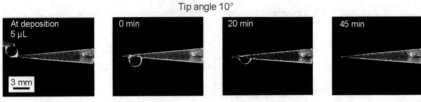

Droplet evaporates in about 45 min.

Fig. 4.13 Images showing the effect of evaporation of a 5 μL droplet placed on the tip of a 10° tip angle cone. The droplet completely evaporated after about 45 min (adapted from Schriner and Bhushan 2019)

centerline in the horizontal direction. Droplets were fed in increments of 5 μL volume to a total volume of 40 μL.

On a cone with the sideline in the horizontal direction, only Laplace pressure gradient due to curvature gradient drives the droplet. Whereas, on a cone with the centerline in the horizontal direction, in addition to Laplace pressure gradient, the gravitational forces due to inclination of the cone surface with respect to horizontal axis, drive the droplet.

Figure 4.14a shows the relationship between droplet volume and distance traveled by the droplet for cones of two tip angles, 10° and 45°, with sideline in horizontal orientation (Gurera and Bhushan 2019b). Figure 4.14c shows the four images, (top) during the moment of the first droplet deposition and just after deposition, and (bottom) equilibrium stages of a 10 μL droplet and a 40 μL droplet, for cones of 10° tip angle (Gurera and Bhushan 2019a).

Figure 4.14a shows that for any cone, a droplet will move towards the base if the droplet volume is increased. On increasing the volume, a droplet instantaneously moves a certain distance and then stops. The droplet moves due to the Laplace pressure gradient resulting from the underlying curvature gradient. As the curvature decreases, the Laplace pressure inside the droplet decreases. The droplet stops because the force due to the Laplace pressure becomes less than the adhesion force between the surface and the droplet. Figure 4.14a also shows that a droplet of known volume travels farther on a cone of tip angle of 10°, as compared to on a cone of tip angle of 45°.

The curvature gradient was calculated using the expression, $(1/R_1 - 1/R_2)$, where R_1 and R_2 are the two local radii of the cone at the front and rear contact lines of the droplet siting on a cone, respectively. As the droplet moves, both R_1 and R_2 change. For a cone with a tip angle of θ, radius, R, at a given distance from the tip, d, is given by $d \tan(\theta/2)$. The curvature gradient was plotted as a function of distance from the tip of the cone to the center of the droplet, along the cone axis. Two droplets with lengths of 0.5 and 2 mm measured along sideline of cones, were selected because they are typical large droplet lengths observed in the water collection measurements. Figure 4.15 shows calculated curvature gradient as a function of distance for two cones with tip angles of 10° and 45°. An increase in the

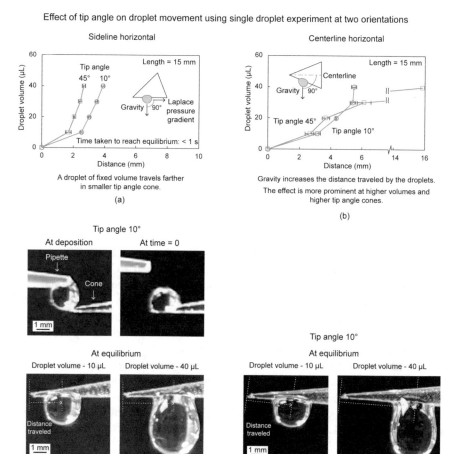

Fig. 4.14 Droplet volume as a function of the distance travelled for cones with 10° and 45° tip angles, when **a** a cone placed with the sideline in horizontal orientation, as shown in the insert, to study the effect of Laplace pressure gradient and to remove the gravitational force in driving the droplet, and **b** a cone placed with the centerline in the horizontal orientation, as shown in the insert, to add gravitational effects in driving the droplet. **c** The four images shown are (top) during the moment of the first droplet deposition and just after deposition, and (bottom) equilibrium stages of 10 and 40 µL droplets on a 10° tip angle cone with sideline in horizontal orientation, and **d** the two images shown at equilibrium stages of 10 and a 40 µL droplets on a 10° tip angle cone with centerline in horizontal orientation (adapted from Gurera and Bhushan 2019b)

droplet length increases the curvature gradient because of the larger change in the radii. The curvature gradient decreases with the distance from the tip. The curvature gradient of a cone with a smaller tip angle is larger and remains so for a longer distance. Therefore, the droplet on a cone with smaller tip angle travels farther. Since curvature gradient decreases with distance, the droplet stops after some distance.

Fig. 4.15 Calculated curvature gradient as a function of distance from the tip of two cones with tip angles of 10° and 45° for two droplets with lengths of 0.5 and 2 mm

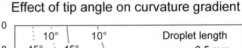

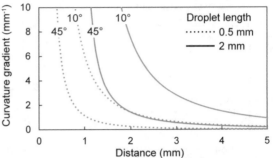

Figure 4.14b shows the relationship between droplet volume and distance traveled by the droplet for cones of two tip angles, 10° and 45°, with centerline in horizontal orientation. Figure 4.14d shows the two images at equilibrium stages of a 10 µL droplet and a 40 µL droplet, for cones of 10° tip angle (Gurera and Bhushan 2019b). Droplets on both cones traveled farther, compared to that of cones with sideline in the horizontal direction. A droplet of 40 µL on 45° tip angle cone traveled the entire cone length and reached its base.

At centerline in the horizontal orientation, a component of gravity along the side of the cone also drives the droplets. The gravitational force acting on the droplets is given by $(V\rho)g \sin(\theta/2)$, where V is volume of the droplet, ρ is mass density of water and g is the gravitational constant (9.8 m s^{-2}). It increases the distance traveled by the droplets along the cone. The effect is more pronounced for larger volume droplets and larger tip angles.

Schriner and Bhushan (2019) also conducted single droplet experiments in the presence of fog. Fog continuously deposits droplets on the entire cone. These droplets coalesce with each other and the increased volume of the droplet increases the distance traveled. It is the continuous deposition of fog droplets which maintains the droplet motion. Otherwise, in the absence of fog deposition, any deposited droplets only travel a certain distance during which Laplace force is high enough to overcome adhesion to the surface.

To summarize, in a single droplet test for a cone with sideline in horizontal orientation, lower tip angle transports a water droplet farther along the cone. This is because, for a lower tip angle, curvature gradient is larger. For a cone in centerline horizontal orientation, gravity increases the distance traveled by droplets. Water droplets travel farther along the cone as compared to that with sideline in the horizontal orientation. To further sum up, for efficient water transport, it is important to maximize the effects of Laplace pressure gradient, gravity and droplet coalescence.

4.2.5 Cones—Effect of Geometry, Inclination, Grooves and Heterogeneous Wettability

Two sets of cones were tested to evaluate the effect of tip angles—one with the same surface area and another with the same length at two inclination angles. Cones with grooves and heterogeneous wettability were also studied. Next, a conical array was tested. Finally, nonlinear cones were tested.

4.2.5.1 Two Tip Angles with Same Surface Area

Figure 4.16 shows water collection rate for cones with two tip angles and with the same surface areas at 0° inclination (Gurera and Bhushan 2019b). The left bar chart shows the water collection rate (mg h^{-1}). The right bar chart shows the average droplet mass (mg) and frequency (droplet h^{-1}) for the two cones. The cone with higher tip angle had higher water collection rate. This is because cone length of the larger tip angle cone is shorter and droplets take lesser time to reach to the base. Although the mass of the falling droplets is similar, the cone with a larger tip angle has a higher frequency of falling droplets.

Figure 4.17 shows a sequence of optical images of water droplets moving from tip to base for both tip angles (Gurera and Bhushan 2019b). A sequence for the first droplet is presented. It was reported that the first droplet takes a longer time to fall as compared to the subsequent droplets. Optical images of water droplets just before they completely detached from the cones are also shown. The size of the droplets appears to be similar, independent of the tip angle. This is the reason that the mass

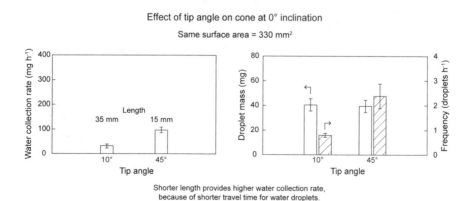

Fig. 4.16 Water collection rates, and mass and frequency of falling droplets for cones with the same surface area and tip angles of 10° and 45°, at 0° inclination angle (adapted from Gurera and Bhushan 2019b)

Droplet movement on cone at 0° inclination

First falling droplet

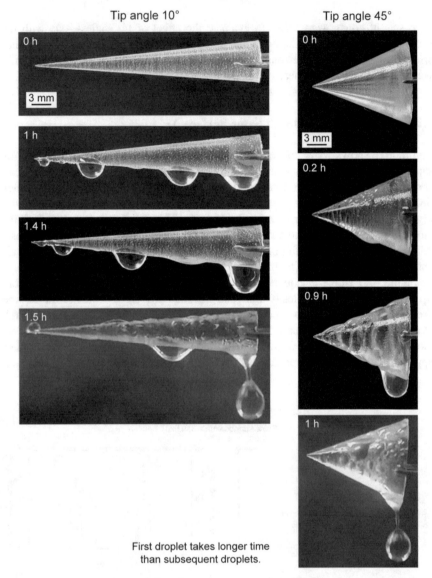

First droplet takes longer time
than subsequent droplets.

Fig. 4.17 Optical images showing movement of water droplets from tip to base on the 10° and 45° tip angle cones and droplets just before and after they detach from the surface and fall. The sizes of the falling droplets on the two cones appear to be similar. It was reported that the first droplet takes a longer time to fall as compared to the subsequent droplets (adapted from Gurera and Bhushan 2019b)

of the fallen droplets was found to be similar. As the droplet starts from the tip and reaches the end of the cone, it sits there and elongates. As the elongation reaches a critical length, the droplet starts to break away as shown in the images.

4.2.5.2   Two Tip Angles with Same Length

Figure 4.18a shows water collection rates for cones with two tip angles of the same length at 0° inclination (Gurera and Bhushan 2019b). The bar chart on the left side shows the water collection rate data (mg h^{-1}). The bar chart on the right shows the average droplet mass (mg) and frequency (droplet h^{-1}) from the two cones. The cones of same length were found to have similar water collection rate. From the single droplet experiment, Laplace pressure gradient was found to be more effective in a smaller tip angle cone. However, due to the difference in surface areas, more water droplets are formed on the surface of the 45° tip angle cone. This makes the water collection rate for both the cones similar. Droplet mass and frequency were reported to be similar for the two cones.

Figure 4.18b shows a sequence of optical images of first and second water droplets moving from tip to base for the cone with 10° tip angle (Gurera and Bhushan 2019b). The first droplet takes the longer time to fall as compared to the subsequent droplets.

4.2.5.3   Inclination Angle

Figure 4.19a shows water collection rates for cones with two tip angles at two inclination angles (Gurera and Bhushan 2019b). The cones of 10° and 45° tip angles, having either the same surface area or same length were compared at two inclination angles of 0° and 45°.

As it has been mentioned before, there are two forces driving droplets on the conical surfaces due to Laplace pressure gradient and gravity. In the previous section, it was established that in the absence of driving gravitational force, at 0° inclination angle, the cone length dictates the water collection rate. The shorter the cone, the higher the water collection rate, irrespective of the tip angle. This is because the distance traveled by the droplets from the tip to the base is shorter in shorter cones.

At a 45° inclination angle, gravity also drives droplets. In addition to role of Laplace pressure gradient and gravity, distribution of droplets on the cone surface also play a role. Figure 4.19b shows the differences in distribution of droplet on the surface area of a 10° tip angle cone at two inclinations. At 0° inclination angle, as indicated earlier, larger droplets were observed near the tip because of higher fog interception. These droplets grow further and move towards the base which takes time. At 45° inclination angle, water droplets were observed on the entire surface area. More uniform distribution will increase the water collection rate. Therefore, increasing the inclination angle increases the water collection rate, irrespective of

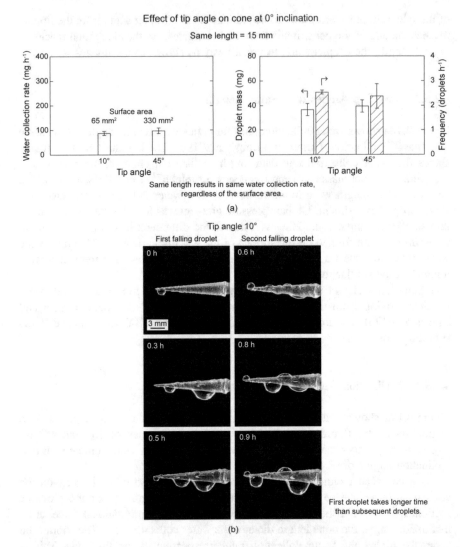

Fig. 4.18 a Water collection rates, and mass and frequency of falling droplets for cones with the same length and tip angles of 10° and 45°, at 0° inclination angle. **b** Optical images showing movement of water droplets from tip to base on a cone with tip angle of 10°. The first droplet took longer time to fall as compared to the subsequent droplets (adapted from Gurera and Bhushan 2019b)

the cone tip angle. The increase is more for cones with a larger surface area. At a higher inclination angle, it is the number of droplets formed on the surface that dictates the water collection rate, not the length.

Effect of inclination angle on single cones with various lengths and surface areas

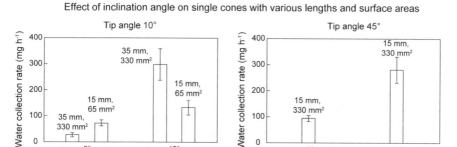

At 0° inclination, length dictates water collection rate.
At 45° inclination, surface area dictates water collection rate.

(a)

Effect of inclination angle on droplet formation

Tip angle 10°, 35 mm, 330 mm² (at 0.1 h)

At 0° inclination   At 45° inclination

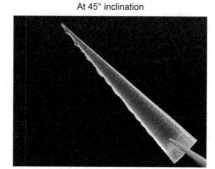

Large droplets are observed at the cone tip   Large droplets are observed on the entire surface
because of fog interception   because of fog interception

(b)

Fig. 4.19 **a** Summary of water collection rates for cones of tip angles of 10° and 45°, with either the same length or the same surface area, and at two inclination angles of 0° and 45° (adapted from Gurera and Bhushan 2019b). **b** Optical images of water droplets for cones of 10° tip angle at two inclination angles of 0° and 45° showing the differences in distribution of droplets at two inclination angles

4.2.5.4 Velocity of Droplets

High droplet velocity is needed for water collection purposes to minimize water loss to evaporation. For velocity measurements, the distance between the center of the droplet from the tip of the cone was measured as a function of time. The center of the droplet was located by taking the midpoint of the distance between the left edge and the right edge of the droplet. Measurements were performed by using a video camera to record a video that started when the second droplet was deposited and ended when the second droplet reached the cone base or detached from the

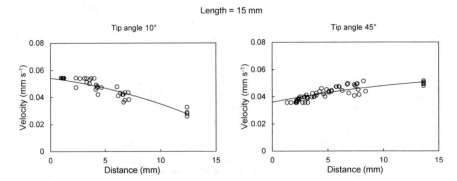

Velocity at tip of the 10° tip angle cone is higher because of larger Laplace pressure gradient.
Velocity decreases with length on 10° tip angle cone because of decreasing curvature gradient.
However, velocity increases with length on 45° tip angle cone because gravity dominates.

Fig. 4.20 Velocity of droplets on cones with 10° and 45° tip angle, at 0° inclination angle as a function of distance from the tip over the length of cones (adapted from Schriner and Bhushan 2019)

cone. Screenshots were taken with timestamps to document droplet movement. These screenshots were then analyzed by software included with the video camera to accurately determine droplet distance on the cone at certain periods of time. These time/distance datapoints were plotted and a second-order polynomial equation trendline was fitted to the data. The derivative of this time/distance equation was then calculated to determine the velocity/time equation. The droplet velocity at any length on the cone could be found by entering the time associated with a distance measurement into the velocity equation (Schriner and Bhushan 2019).

Velocities of droplets as a function of distance moving on cones of 10° and 45° tip angles, at 0° inclination are shown in Fig. 4.20 (Schriner and Bhushan 2019). The initial velocity of the cone with 10° tip angle was higher than that of the cone with a 45° tip angle because of larger Laplace pressure gradient. For the 10° tip angle cone, the velocity decreases with distance because of the decreasing curvature gradient responsible for the Laplace pressure gradient. However, for the 45° tip angle cone, velocity does not decrease with distance because gravity plays a larger role. This cone also benefits from larger surface area because of the additional water droplets formed on its whole surface (Schriner and Bhushan 2019).

4.2.5.5 Grooves

To study the effect of grooves, a representative cone with a tip angle of 45° and a length of about 15 mm was selected. Figure 4.21a shows the water collection rates (mg h^{-1}) of ungrooved and grooved cones at two inclination angles, 0° and 45° (Gurera and Bhushan 2019b). At a 0° inclination angle, grooves help in increasing

Fig. 4.21 a Water collection rates for ungrooved and grooved cones with tip angle of 45°, at 0° and 45° inclination angles. **b** A water droplet of volume of 5 μL was placed on ungrooved and grooved cones to demonstrate the channeling in the grooved cone (adapted from Gurera and Bhushan 2019b)

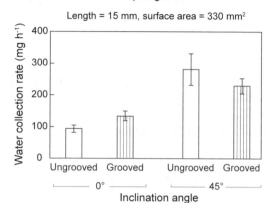

Effect of grooved cone

Tip angle 45°

Length = 15 mm, surface area = 330 mm²

At 0° inclination, grooves contribute.
At 45° inclination, gravity dominates and grooves do not contribute.

(a)

Channeling of water droplets

0° inclination

(b)

the water collection rate. This is because grooves help in channeling the water. It is believed that they also create an increased Laplace pressure gradient via the gradient in grooves spacing. However, at a 45° inclination angle, the improvement due to grooves is not observed. That is probably because the gravitational force is overpowering the advantage of grooves.

To study the differences in the channeling of a water droplet on the ungrooved and grooved cones, a fixed volume of water droplet, 5 μL, was placed on the cones at similar locations (Gurera and Bhushan 2019b). The optical images are shown in Fig. 4.21b. The droplets between the grooves appears to be more channeled towards the base. This elongation of the droplet is the reason for the increase in the water collection rate in the grooved cone.

4.2.5.6 Heterogeneous Wettability

To study the effect of heterogeneous wettability, a cone with a tip angle of 45° and a length of about 15 mm was selected. Figure 4.22 shows the effect of heterogeneity on water collection rate (mg h^{-1}) at two inclination angles of 0° and 45° (Gurera and Bhushan 2019b). The water collection rates are compared with the cones with heterogeneous wettability. At 0° inclination angle, the heterogeneity increases the water collection rate. This is because the heterogeneity assists in transporting the water quickly. However, at a 45° inclination angle, an improvement due to heterogeneity is not observed. This is probably because the gravitational force dominates the transport and not the heterogeneity.

For multistep wettability gradient studies, see Gurera and Bhushan (2020).

Fig. 4.22 Water collection rates for cones with a tip angle of 45° and homogeneous and heterogeneous wettability, and at 0° and 45° inclination angles (adapted from Gurera and Bhushan 2019b)

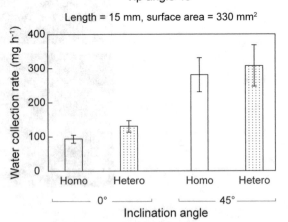

4.2.5.7   Conical Array

An array with cones with a tip angle of 45° and a length of about 15 mm were selected to evaluate effects of arrays at two inclination angles of 0° and 45°. Figure 4.23a shows a water collection rate (mg h^{-1}) for an array (Gurera and

Fig. 4.23 a Water collection rate of a single cone and an array at 0° and 45° inclination angles. The array data is presented as water collection rate per cone. **b** Optical images of falling droplets on array at 0° inclination angle showing coalescence (adapted from Gurera and Bhushan 2019b)

Droplets falling from the top cone coalesce with droplets stuck to cone underneath.

(b)

Bhushan 2019b). The array data is presented as water collection rate per number of cones. The data are compared with a single cone with a tip angle of 45°. At a 0° inclination angle, having an array increased the water collection rate per number of cones. It is believed that this increase is due to a cascading effect on a falling droplet. This is because a droplet falling from the top cone collects the droplets stuck to the cone underneath increasing the net water collection rate per cone, as shown in Fig. 4.23b (Gurera and Bhushan 2019b). At a 45° inclination angle, the data for a single cone and the array per cone is comparable. Water collection rates at 45° inclination angle is larger than at 0° inclination angles. As suggested earlier, this occurs because the gravitational forces dominate the water transport at a higher inclination.

4.2.5.8 Summary

For water collection on conical surfaces from fog, water droplets are driven by the Laplace pressure gradient and gravity. During travel, droplets coalesce and travel farther because of larger mass. The Laplace pressure gradient dominates at 0° inclination angle, which results in higher water collection rate for shorter length. Gravity dominates the water collection rate at 45° inclination angle, which results in higher water collection rate for larger surface area. The water collection rate remains independent of the tip angle for same length but larger at larger tip angle for same surface area, irrespective of the inclination angle. The water collection rate always increases with an increase in inclination angle, regardless of the cone, because of gravity effects. Grooves and heterogeneous wettability also increase the water collection rate. In arrays, water collection per cone is higher at 0° inclination angle than a single cone. However, it is similar at 45° inclination (Gurera and Bhushan 2019b).

4.2.6 Nonlinear Cones

Nonlinear profile of a cone can be used to increase the curvature gradient near the tip and higher slope near the base, in order to maximize water collection for a cone with a given length and base area. Gurera and Bhushan (2019c) used a cone with a length of 15 mm and base diameter of 13 mm with a concave profile with low tip angle of 10° to increase the Laplace pressure gradient near the tip and nonlinearly to increase the radius to have higher gravitational forces in the higher slope region.

The curvature gradient was calculated using the expression, $(1/R_1 - 1/R_2)$, where R_1 and R_2 are the two local radii of the cone at the front and rear contact lines of the droplet sitting on a cone, respectively. As the droplet moves, both R_1 and R_2 change.

Figure 4.24a shows variation of curvature gradients as a function of distance from the tip, on linear cones of tip angle, 10° and 45°, and the nonlinear cone, for a

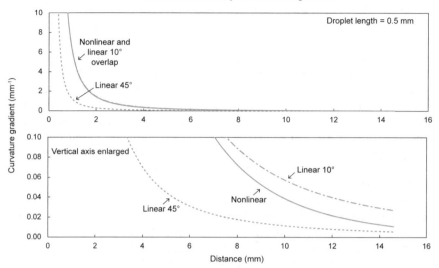

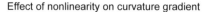

Curvature gradient of the nonlinear cone is similar to the 10° tip angle linear cone near the tip.
Curvature gradient of the nonlinear cone converges to the 45° tip angle linear cone near the base.

(a)

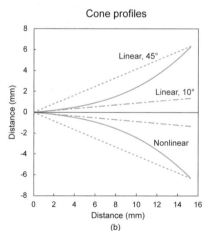

(b)

Fig. 4.24 **a** Effect of the nonlinearity on the calculated curvature gradient as a function of distance from the tip of linear and nonlinear cones, using a droplet length of 0.5 mm. The top graph shows the curvature gradient variation near the tip and the bottom graph shows the variation near the base with magnified vertical axis. On the linear cones, the curvature gradient is higher with shorter tip angle. The curvature gradient of the nonlinear cones starts out overlapping with the 10° tip angle linear cone, because of the same tip angle (top graph). Later, the curvature gradient remains between 10° and 45° tip angle cones and converges towards the 45° tip angle cone (bottom graph). **b** Overlapped cones' profile for the linear and the nonlinear cones (adapted from Gurera and Bhushan 2019c)

droplet length of 0.5 mm. The top graph shows the curvature gradient variation near the tip and the bottom graph shows the variation near the base with magnified vertical axis. Figure 4.24b shows overlapped profiles of the three cones for reference. The following observations can be made. The curvature gradient decreases with length because of the decreasing curvature of the cones. As reported earlier, the curvature gradient of the 10° tip angle cone is higher than the 45° tip angle cone because of its radius increasing at a slower rate. Curvature gradient of the nonlinear cone was similar to the linear, 10° tip angle cone for the initial cone length. For the latter part of the cone, the curvature gradient of the nonlinear cone becomes between the linear, 10° and 45° tip angle cones and converges towards the linear, 45° tip angle cone with the same base diameter. It is expected that because of the increasing slope of the cone, the gradient should be converging towards the higher tip angle cone.

Both single droplet and water collection from fog experiments were performed. Figure 4.25 shows the relationship between droplet volume and distance traveled by the droplet for the linear and the nonlinear cones. The nonlinear cone transports the liquid father than the linear cones. This is because of the high Laplace pressure gradient in the beginning and increasing gravitational effect because of higher slope later.

Figure 4.26 shows the water collection data for linear and nonlinear cones. Figure 4.26a shows the water collection rate (mg h^{-1}) and Fig. 4.26b shows average droplet mass (mg) and frequency (droplet h^{-1}) for the cones. The water collection rate for the linear cones were similar (Gurera and Bhushan 2019b), however, the water collection rate of the nonlinear cone was found to be higher.

Fig. 4.25 Effect of tip angle and the nonlinearity on the droplet movement along the cones in single droplet experiments. The cones were placed with the centerline in the horizontal direction, as shown in the schematic. The nonlinear cone transports the water droplets farthest, among the three cones. This is because of the high Laplace pressure gradient in the beginning and increasing gravitational forces later (adapted from Gurera and Bhushan 2019c)

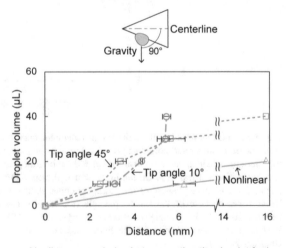

Effect of tip angle and nonlinearity on droplet movement using single droplet experiment

Nonlinear cone helps in transporting the droplet farther.

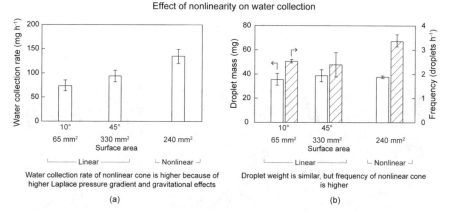

Fig. 4.26 Effect of nonlinearity on water collection for the linear and the nonlinear cones. **a** Water collection rates (mg h^{-1}), and **b** droplet mass (mg) and the frequency (droplets h^{-1}). The water collection rate and frequency of the nonlinear cone was found to be highest, because of high Laplace pressure gradient in the beginning and increasing gravitational forces later. Droplet mass remained similar regardless of the cone (adapted from Gurera and Bhushan 2019c)

It is believed that because of high Laplace pressure gradient in the beginning and increasing gravitational effects later, the droplets were transported faster, which was observed in the higher frequency of the falling droplets. The mass of the falling droplets remained similar.

Figure 4.27 shows a sequence of optical images of water droplets traveling from tip to base for the linear and the nonlinear cones. Only the second droplet has been shown in the figure, to present a representative time taken by each cone to form a droplet big enough to fall. The linear cones take similar time and have similar hanging droplet size. The nonlinear cone takes lesser time and has similar size of the hanging droplet. These observations are in agreement with the droplet mass and frequency measurements, discussed earlier.

To sum up, a nonlinear cone provides higher water collection from fog than that for a linear cone with 10° tip angle and another cone with the base diameter of 13 mm. The nonlinear cone was designed with small tip angle of 10° to increase the Laplace pressure gradient near the tip due to higher curvature gradient and non-linearly increasing radius to have higher gravitational effects from the higher slopes. Therefore, for a high water collection rate, a smaller tip angle and broader base diameter with higher surface slope is desired.

4.2.7 Projection for Water Collection Rates

In fog deposition at 0° inclination angle (samples place horizontally), water droplets will get deposited on flat surfaces with any wettability, but will not be collected in

Droplets traveling on linear and nonlinear cones

(After first droplet has fallen)

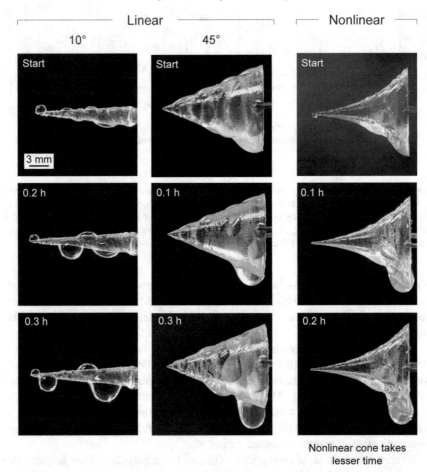

Fig. 4.27 Optical images showing water droplets traveling from tip to base for the linear and the nonlinear cones. Only the second droplet has been shown, to present a representative time taken by each cone to form a droplet big enough to fall. The linear cones take similar time and have similar hanging droplet size. The nonlinear cone takes lesser time and have similar size of the hanging droplet (adapted from Gurera and Bhushan 2019c)

the reservoir. In the case of hydrophilic cones, water collection occurs in the reservoir (near their bases), and the data is shown in Figs. 4.10 and 4.19. It is also expected, in the case of beetle-inspired surfaces, some water collection would occur.

At 45° inclination angle, based on Figs. 4.10, 4.12, and 4.19, water collection rates for a hydrophilic cone at 45° inclination angle is more than factor of 10 larger than at 0° inclination angle. Water collection rate is larger than that at 0° inclination angle because of the effect of gravity.

At 45° inclination angle, based on Fig. 4.9, for beetle-inspired surfaces and hydrophilic cones, water collection rates per unit surface area is about 1 mg mm^{-2} h^{-1}, which is about twice as much water as that for a hydrophilic flat surface. Based on Figs. 4.21, 4.22 and 4.23, cones with grooves and heterogeneous wettability and conical arrays provide slightly larger water collection. Furthermore, collected water is transported at a faster rate on a beetle-inspired and cone surfaces than that on a flat surface which reduces the evaporation of water droplets before reaching storage/use location. It was reported earlier, based on Fig. 4.13, that water droplet can evaporate in about 45 min. Finally, based on Fig. 4.19, at 45° inclination angle, water collection rates are larger for larger surface area for both cone angles. For the data shown, an increase of 3–6 times can be achieved. Therefore, cones of larger surface area for the same base area should be used.

Based on these observations, for a conical array with grooves and heterogeneity, water collection rates per unit base area as compared to that of flat hydrophilic surface can be increased by about a factor of 10 or more. In addition, water is transported faster which reduces evaporation. A summary of projected collection rates is presented in Table 4.2 (Bhushan 2020).

For completeness, we compare typical water collection rates from fog in the desert with that obtained in the lab. Typical water collection rate in the desert, based on Figs. 2.7 and 2.8 is on the order of 2 L m^{-2} day^{-1} (or 2 mg mm^{-2} day^{-1}). Hourly collection rate is about 1 mg mm^{-2} h^{-1}. In the lab, based on Fig. 4.12, water collected on a flat hydrophilic surface is about 0.5 mg mm^{-2} h^{-1}. Therefore, for the experimental conditions used in the lab, water collection rate is about five times that in the desert.

Table 4.2 Summary of projected water collection rates from fog (adapted from Bhushan 2020)

Surfaces	Increase in projected water collection rates
0° inclination angle (horizontal)	
Flat surfaces with any wettability	No collection in the reservoir
Beetle-inspired surfaces and hydrophilic cones	Finite collection
45° inclination angle	
Beetle-inspired surfaces and hydrophilic cones relative to hydrophilic flat surface	2×
Increase in surface area of cones for the same base area	>3×
Cones with grooves and/or heterogeneous wettability	>1×
Conical array per cone relative to a single cone	>1×
Total	∼10×

The water collection rates at 45° inclination angle are about an order of magnitude larger than that at 0° inclination angle

4.3   Results and Discussion—Water Condensation Studies

Water collection data from condensation on cones are presented in this section. First, data on a cylinder and a cone are presented to demonstrate the benefit of conical shape. Next, data on cones with two tip angles having either the same surface area or the same length are presented. This is followed by data at two inclination angles and array. Finally, trends observed in water collection from condensation have been discussed with respect to the trends observed in water collection from fog (Gurera and Bhushan 2020).

4.3.1   Cylinder Versus Cone

In nature, the collected water that matters is which reaches the base. Conical shapes help to transport water droplets towards the base by developing a Laplace pressure gradient in droplets. To demonstrate the role of Laplace pressure gradient, water collection data for cylinder and cone were compared. A representative cylinder and cone were chosen with 35 mm length and 330 mm^2 surface area, and 10° cone tip angle. To minimize any gravitational effects, cylinder and cone were inclined at 0°.

Figure 4.28 shows water collection versus time for the cone and cylinder (Gurera and Bhushan 2020). Since the water that matters is which collects near the base, the water was collected from half-length at base. As mentioned earlier, the slope of the fitted straight line is referred to as the water collection rate (mg h^{-1}). Figure 4.29a presents comparison of water collection rate of cylinder and cone, for complete collection versus half-length collection at base (Gurera and Bhushan 2020). The left bar chart presents the water collection rate (mg h^{-1}). The right bar chart presents average droplet mass (mg) and frequency (droplet h^{-1}) for the surfaces (Gurera and Bhushan 2020). The left half of each bar chart represents water

Fig. 4.28 A representative water collection (mg) versus time (h) plot for hydrophilic cone and cylinder surface with 35 mm length and 330 mm^2 surface area, at 45° inclination angle. The cone had tip angle with 10°. The slope of the curve determines their water collection rate (mg h^{-1}) (adapted from Gurera and Bhushan 2020)

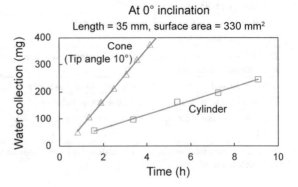

Water collection at base versus time for cone versus cylinder

At 0° inclination

Length = 35 mm, surface area = 330 mm^2

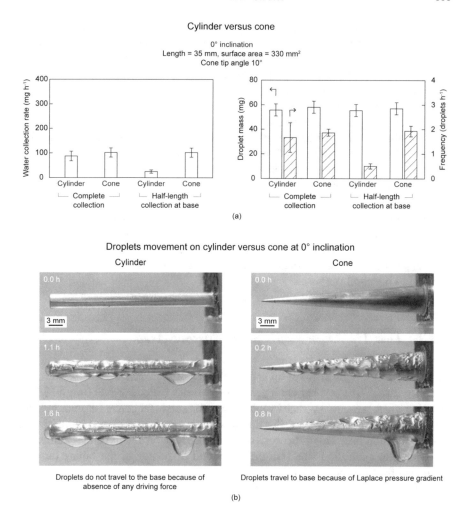

Fig. 4.29 **a** Water collection rates and mass and frequency of falling droplets of cylinder and cones. The left half of both bar charts shows the data when water collected under the entire surface, that is complete collection. The right half of the bar charts shows the data when water was collected under the half-length near base, as in nature. **b** The optical images showing movement of water droplets on a cylinder and a cone (adapted from Gurera and Bhushan 2020)

collection rate for complete collection and the right half represents water collection rate for half-length collection at base. Figure 4.29b presents droplets movement on cylinder (left) and cone (right) (Gurera and Bhushan 2020). The final time stamp represents time taken by the first droplet to fall.

When comparing cylinder and cone for complete collection, both result in similar water collection rates, Fig. 4.29a. This is because same surface area results in similar number of water droplet nucleation sites. Droplet mass and frequency of the falling droplets were also similar. Droplet mass is believed to be a result of various factors including shape, size and wettability of the surface. Frequency is believed to be a result of the rate at which vapors were condensed on the surface. Product of these two result to be the water collection rate. Therefore, similar droplet mass and similar frequency of falling droplets result in similar water collection rate.

When looking at cylinder there is a difference in complete collection versus half-length collection at base, Fig. 4.29a. This is because there were multiple droplets on the cylinder's surface at various locations and typically only one droplet falls into the beaker at the half-length location, as shown in optical images showing movement of droplets in Fig. 4.29b. Typically, three droplets fall when looking at complete collection. That results in lowering of the water collection rate of cylinder for half-length collection at base. There is an absence of any droplet transporting forces such as Laplace pressure gradient.

When looking at cone, there is no difference between complete collection versus half-length collection at base, Fig. 4.29a. This is because droplets travel from tip to base due to Laplace pressure gradient, Fig. 4.29b. Therefore, conical shapes are advantageous, are typically found in nature for water transport, and should be used for water collection purposes.

4.3.2   Two Tip Angles with Same Surface Area

Figure 4.30 shows the effect of tip angle on cones of same surface area (Gurera and Bhushan 2020). The left bar chart presents the water collection rate (mg h^{-1}). The right bar chart presents average droplet mass (mg) and frequency (droplet h^{-1}) for the two cones. Irrespective of the tip angle, same surface area results in similar water collection rate. This is because same surface area results in similar number of water droplet nucleation sites. Droplet mass and frequency of the falling droplets were also found to be similar.

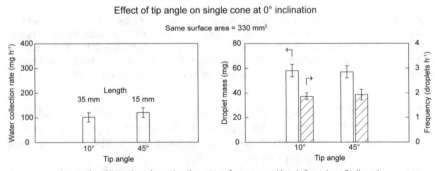

Fig. 4.30 Water collection rates and mass and frequency of falling droplets of cones with the same surface area and different tip angles, 10° and 45°, compared at 0° inclination angle. The water collection rates of either of the cones were found to be similar, and so were mass and frequency of the falling droplets (adapted from Gurera and Bhushan 2020)

Figure 4.31 shows a sequence of optical images of water droplets traveling from tip to base for both of the tip angles (Gurera and Bhushan 2020). A sequence for the first and the second droplet is presented. The first droplet takes the longest time to

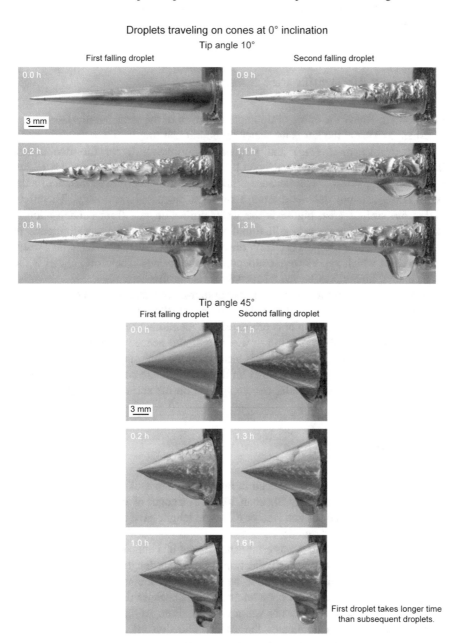

Fig. 4.31 Optical images showing water droplets traveling from tip to base on cones of two tip angles, 10° and 45°, and surface area of 330 mm². The time taken for second and following droplets was similar in both cases (adapted from Gurera and Bhushan 2020)

fall as compared to the subsequent droplets. And as mentioned before, therefore, the time taken for the first droplet to fall have been ignored from all the data and calculations shown in this study. It is clear from the time steps of the second droplet that the time taken for it to fall is similar on either of the cones. And similar applies to subsequent droplets.

4.3.3   Two Tip Angles with Same Length

Figure 4.32a shows the effect of tip angle on cones of the same length (Gurera and Bhushan 2020). On the left is the water collection rate (mg h^{-1}). On the right is the average droplet mass (mg) and frequency (droplet h^{-1}) for the two cones. Irrespective of the tip angle and cone length, larger surface area provides higher water collection rates. This is because condensation droplets were formed across the entire cone, and not deposited from a flow, as discussed previously. Therefore, higher the surface area, higher is the water collection rate from condensation is expected. Droplet mass was still found to be similar. However, a difference in frequency of falling droplets is clearly observed, because it is believed to be a result of the rate at which vapors were condensed on the surface. Therefore, higher is the surface area, higher will be the frequency of falling droplets.

Figure 4.32b shows a sequence of optical images of water droplets moving from tip to base for the cone of 10° tip angle (Gurera and Bhushan 2020). A sequence for the first and the second droplet is presented. As mentioned before, here as well, the time taken by the first droplet to fall was longer and hence has been ignored from this study. For the second droplet, it is clear from the time steps that the time taken by it to fall is longer than the droplet on the 45° tip angle cone of same length, as shown in Fig. 4.31. And similar can be applied on the subsequent droplets.

4.3.4   Inclination Angle

Figure 4.33 shows the water collection rate of the cones of two tip angles at two inclination angles (Gurera and Bhushan 2020). The cones of 10° and 45° tip angles, having either the same surface area or the same length have been compared at two inclination angles of 0° and 45°.

In the previous section, it was established that at 0° inclination angle, the surface area of cone dictates the water collection rates. Cone length did not have a significant effect. Higher is the surface area, higher is the water collection rates, irrespective of the tip angle or cone length. This is because the droplets are formed across the entire surface by water vapor condensation. At 45° inclination angle as well, the surface area dominates the water collection rate, because the droplet formation is still independent of the inclination angle.

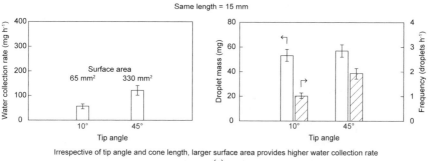

Effect of tip angle on single cone at 0° inclination

Same length = 15 mm

Irrespective of tip angle and cone length, larger surface area provides higher water collection rate

(a)

Tip angle 10°

First falling droplet   Second falling droplet

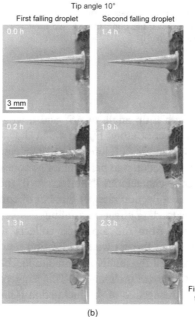

First droplet takes longer time than subsequent droplets.

(b)

Fig. 4.32 a Water collection rates and mass and frequency of falling droplets of cones with the same length and different tip angles, 10° and 45°, are compared at 0° inclination angle. The water collection rates of 10° tip angle cone was found to be lower, because it has lower surface area. Mass of the falling droplets was found to be similar, however, frequency of the falling droplets was found to be lower for the 10° tip angle cone. **b** Optical images showing water droplets traveling from tip to base on cone of 10° tip angle and surface area of 65 mm^2. The time taken for the second droplet and the following droplets for the 10° tip angle cone was longer (adapted from Gurera and Bhushan 2020)

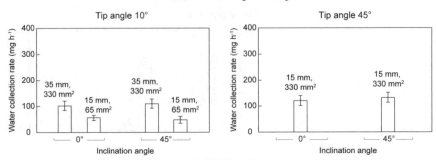

Fig. 4.33 Water collection rates of cones of two tip angles, 10° and 45°, with either the same lengths or the same surface areas, are compared at two inclination angles: 0° and 45° (adapted from Gurera and Bhushan 2020)

4.3.5 Array

Representative cones with a tip angle of 45° and a length of about 15 mm were chosen to evaluate the effect of array. Figure 4.34a shows the effect of an array on the water collection rate (mg h^{-1}) (Gurera and Bhushan 2020). The array data is presented as the water collection rate per number of cones. Two inclination angles, 0° and 45° were chosen to characterize the effect. At 0° inclination angle, having an array increases the water collection rate per number of cones. It is believed that this increase was due to a cascading effect on a falling droplet, as shown in Fig. 4.34b. In that a falling droplet from the top cone collects the droplets stuck to the underneath cones increasing the net water collection rate per number of cones. At 45° inclination angle as well the trends remain similar, because of the indifference in droplet formation at different inclinations. This suggests that the cascading does provide an advantage, irrespective of the inclination (Gurera and Bhushan 2020).

4.3.6 Trends in Water Collection in Condensation Versus Fog

In nature, condensation and fog often occur simultaneously, therefore, it is of interest to compare mechanisms of water collection by both. A cone with 10° tip angle, 35 mm length, and 330 mm^2 surface area is selected to demonstrate the differences. Differences can be studied by comparing the initial droplets formation between the two. Figure 4.35 shows differences in droplet formation due to condensation and fog at 0.1 h, at 0° and 45° inclination (Gurera and Bhushan 2020).

Fig. 4.34 **a** The water collection rate of a single cone and an array at 0° and 45° inclination angles. The array data is presented as the water collection rate per cone. Array increases the water collection rate per cone. **b** Optical images of falling droplets on array at 0° inclination angle showing coalescence. The array increases the water collection rate because a falling water droplet will collect the water droplets stuck on the cones underneath (adapted from Gurera and Bhushan 2020)

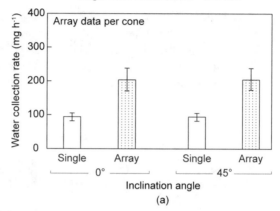

Effect of array

Tip angle 45°

Length = 15 mm,
Single cone surface area = 330 mm²

(a)

Cascading effect in arrays

Droplets falling from the top cone coalesce
with droplets stuck to cone underneath and
get collected fast

(b)

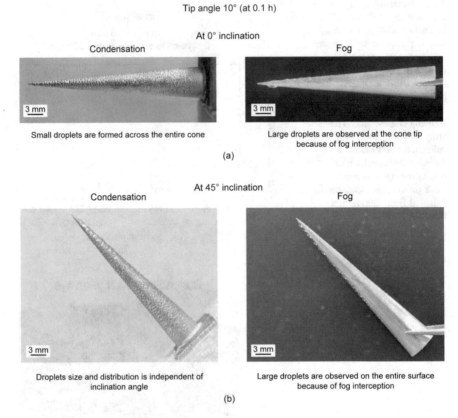

Fig. 4.35 Differences in droplet formation from condensation and fog on a cone with 10° tip angle at **a** 0° inclination angle, and **b** 45° inclination angle (adapted from Gurera and Bhushan 2020)

In condensation, it was reported earlier that at 0° inclination, the water collection rate is dictated by the cone surface area. Higher is the surface area, higher is the water collection rate. In fog, it has been reported that the water collection rate is dictated by cone length (Gurera and Bhushan 2019b; Bhushan 2019a). Shorter is the length, higher is the water collection rate. The difference arises due to the differences in droplet formation, as shown in Fig. 4.35a. In condensation, droplets were observed across the entire surface, because water vapors condense on the entire surface area. The number of droplets increase with increase in surface area. However, in fog, larger droplets were observed near the tip because of higher fog interception. These droplets grow further and move towards the base. It takes longer for droplets to move on a loner cone. Thus, in condensation, surface area plays the

Table 4.3 Summary of water collection rates by condensation and fog for various cones, at two inclination angles

Water source	Inclination angle	Water collection rate (mg h^{-1})		
		10° tip angle; 35 mm length; 330 mm^2 surface area	45° tip angle; 15 mm length; 330 mm^2 surface area	10° tip angle; 15 mm length; 65 mm^2 surface area
Condensation[a]	0°	100	100	50
	45°	100	100	50
Fog[b]	0°	30	100	100
	45°	300	300	130

[a]Gurera and Bhushan (2020)
[b]Gurera and Bhushan (2019b)

role and cone length does not, and in fog, cone length plays the role and surface area does not.

In condensation, it was also reported earlier that at 45° inclination, water collection rate is dictated by the cone surface area. Same trends have been reported in fog (Gurera and Bhushan 2019b; Bhushan 2019a). In both, higher is the surface area, higher is the water collection rate. This is because the droplets were formed on the entire surface in the both cases, Fig. 4.35b.

Finally, the water collection rates for condensation and fog cannot be compared because data was taken using different set of equipment and conditions. The differences may result in different set of results. However, for the chosen conditions, water collection rate by either of the condensation or fog is between 30 and 300 mg h^{-1}. Rates are more comparable at 0° inclination as compared to 45° inclination. The water collection rates are summarized in Table 4.3 (Gurera and Bhushan 2020).

4.4 Design Guidelines for Water Harvesting Systems

Condensation and fog often exists simultaneously in deserts. Fog mostly exists in the nights, and condensation is found during early morning hours. Therefore, to design water harvesting systems both should be taken into account.

In cones, both the Laplace pressure gradient and gravity are the important factors in driving water droplets towards the base. The Laplace pressure gradient dominates at 0° inclination angle (with samples placed horizontally), and shorter cone length will provide higher water collection rate at the base. The gravity dominates at 45° inclination angle, and higher surface area will provide higher water collection rate at the base. Higher inclination angle provides a higher water collection rate, as compared to a lower inclination angle because of the contributions by gravity.

Therefore, for a high water collection rate, a higher surface area, inclined at higher inclination angle is desired. One can increase the surface area which would increase water collection by several fold, an order of magnitude or more. The surface area can be maximized by having a larger number of cones per unit base area and longer cones. The dimensions of the cones are limited by their structural integrity. Grooves, heterogeneous wettability and nonlinear shapes can also increase water collection rates.

It was found that, from fog, both beetle-inspired and conical surfaces provide about twice the water collection rates per unit surface area than that of homogeneous wettable, flat surfaces. For conical arrays with grooves and heterogeneous wettability, the water collection rates per unit base area can be increased by about a factor of 10. In addition, water is transported faster which reduces evaporation.

Fig. 4.36 Schematics of bioinspired water harvesting designs with **a** heterogeneous wettability, and **b** conical array, grooves, and heterogeneous wettability (adapted from Bhushan, 2018)

Bioinspired water harvesting design with heterogeneous wettability

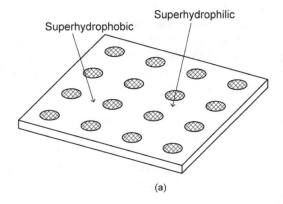

(a)

Bioinspired water harvesting design with conical array, grooves, and heterogeneous wettability

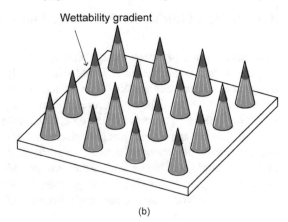

(b)

Fig. 4.37 Schematics of bioinspired water harvesting 2D nets with **a** heterogeneous wettability, and **b** conical array, grooves, and heterogeneous wettability (adapted from Bhushan, 2018)

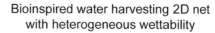

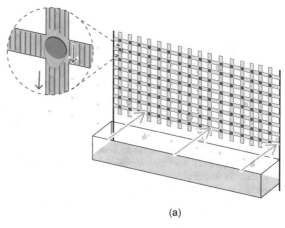

(a)

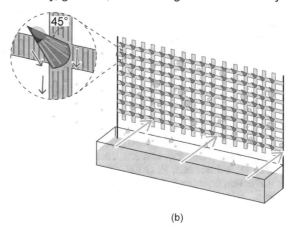

(b)

Design of bioinspired water collectors is shown schematically in Fig. 4.36 (Bhushan, 2018). In one design, surface consists of superhydrophilic spots over a flat superhydrophobic surface. In another design, the surface consists of an array of cones with heterogeneous wettability and grooves. The heterogeneity includes a hydrophobic tip and a superhydrophilic base. The surfaces can be inclined at 45° to the wind for high water collection.

For scaleup, nets or towers can be fabricated using the designs presented in Fig. 4.36. Figure 4.37a and b show 2D water collector nets. Bioinspired water harvesting 3D tower designs for maximum water harvesting will be presented in Chap. 6 (Bhushan, 2020).

References

Azad, M. A. K., Ellerbrok, D., Barthlott, W., and Koch, K. (2015), "Fog Collecting Biomimetic Surfaces: Influence of Microstructure and Wettability," *Bioinspir. Biomim.* **10**, 016004.

Bai, H., Wang, L., Ju, J., Sun, R., Zheng, Y., and Jiang, L. (2014), "Efficient Water Collection on Integrative Bioinspired Surfaces with Star-Shaped Wettability Patterns," *Adv. Mater.* **26**, 5025–5030.

Bhushan, B. (2013), *Introduction to Tribology*, second ed., Wiley, New York.

Bhushan, B. (2017), *Springer Handbook of Nanotechnology*, fourth ed., Springer International, Cham, Switzerland.

Bhushan, B. (2018), *Biomimetics: Bioinspired Hierarchical-Structured Surfaces for Green Science and Technology,* third ed., Springer International, Cham, Switzerland.

Bhushan, B. (2019a), "Bioinspired Water Collection Methods to Supplement Water Supply," *Phil. Trans. R. Soc. A* **377**, 20190119.

Bhushan, B. (2019b), "Lessons from Nature for Green Science and Technology: An Overview and Bioinspired Superliquiphobic/philic Surfaces," *Phil. Trans. R. Soc. A* **377**, 20180274.

Bhushan, B. (2020), "Design of Water Harvesting Towers and Projections for Water Collection from Fog and Condensation," *Phil. Trans. R. Soc. A* **378**, 20190440.

Bhushan, B., and Martin, S. (2018), "Substrate-Independent Superliquiphobic Coatings for Water, Oil, and Surfactant Repellency: An Overview," *J. Colloid Interface Sci.* **526**, 90–105.

Brown, P. S., and Bhushan, B. (2015), "Bioinspired, Roughness-Induced, Water and Oil Super-Philic and Super-Phobic Coatings Prepared by Adaptable Layer-by-Layer Technique," *Sci. Rep.* **5**, 14030.

Garrod, R. P., Harris, L. G., Schofield, W. C. E., McGettrick, J., Ward, L. J., Teare, D. O. H., and Badyal, J. P. S. (2007), "Mimicking a Stenocara Beetle's Back for Microcondensation Using Plasmachemical Patterned Superhydrophobic−Superhydrophilic Surfaces," *Langmuir* **23**, 689–693.

Gurera, D. and Bhushan, B. (2019a), "Designing Bioinspired Surfaces for Water Collection from Fog," *Phil. Trans. R. Soc. A* **377**, 20180269.

Gurera, D. and Bhushan, B. (2019b), "Optimization of Bioinspired Conical Surfaces for Water Collection from Fog," *J. Colloid Interface Sci.* **551**, 26–38.

Gurera, D. and Bhushan, B. (2019c), "Bioinspired Conical Design for Efficient Water Collection from Fog," *Phil. Trans. R. Soc. A* **377**, 20190125.

Gurera, D. and Bhushan, B. (2019d), "Multistep Wettability Gradient on Bioinspired Conical Surfaces for Water Collection from Fog," *Langmuir* **35**, 16944–16947.

Gurera, D. and Bhushan, B. (2020), "Designing Bioinspired Conical Surfaces for Water Collection from Condensation," *J. Colloid Interface Sci.* **560**, 138–148.

Ju, J., Xiao, K., Yao, X., Bai, H., and Jiang, L. (2013), "Bioinspired Conical Copper Wire with Gradient Wettability for Continuous and Efficient Fog Collection," *Adv. Mater.* **25**, 5937–5942.

Schriner, C. T. and Bhushan, B. (2019), "Water Droplet Dynamics on Bioinspired Conical Surfaces," *Phil. Trans. R. Soc. A* **377**, 20190118.

Chapter 5
Bioinspired Triangular Patterns on Flat Surfaces for Water Harvesting

It has been mentioned in Chap. 3, that cactus spines and spider silk take advantage of the conical geometry to drive water droplets by Laplace pressure gradient for water transport and storage/use, before they are evaporated. Bioinspired surfaces with conical geometry for water harvesting from fog and/or condensation of water vapor have been inspired by the cactus spine and spider silk (Bhushan 2018, 2019). In addition, triangular geometries on flat surfaces, inspired by cactus spines, have been pioneered by Song and Bhushan (2019a, b, c, d) for water harvesting from fog and/or condensation of water vapor.

In the case of a cone or triangular geometry, if a droplet is placed at its apex, the droplet is driven across the triangular region by Laplace pressure gradient. The triangular patterns with various geometry and wettability surrounded with a region of less wettability to constrain condensed water droplets inside the patterns have been studied. The wettability of the triangular pattern affects the water collection process and the hydrophilic pattern has been shown to be desirable for water transport compared to the superhydrophilic and hydrophobic patterns (Song and Bhushan 2019a). Hydrophilic triangular patterns of various geometry have been investigated for high water collection rates (Song and Bhushan 2019a, b, c, d).

An important consideration in efficient water collection is to transport water as rapidly as possible. In addition to conical or triangular geometry, droplets can also be driven on surfaces with wettability gradient (Brochard 1989; Chaudhury and Whitesides 1992; Gurera and Bhushan 2019; Feng and Bhushan 2020). When a droplet sits on a surface with heterogeneous wettability, wettability gradient provides an unbalanced force on both sides of the droplet as a driving force for directional transport of the droplet. Multistep wettability gradient has been used on triangular patterns to accelerate droplet transport (Feng and Bhushan 2020). The effects of wettability gradient and triangular geometry on water condensation and transport have been investigated.

One of the limitations of using triangular patterns is that the magnitude of the Laplace pressure gradient decreases along the length of the pattern and the droplet volume required for transport becomes larger. To overcome these limitations,

© Springer Nature Switzerland AG 2020

B. Bhushan, *Bioinspired Water Harvesting, Purification,*
and Oil-Water Separation, Springer Series in Materials Science 299,
https://doi.org/10.1007/978-3-030-42132-8_5

Bhushan and Feng (2020) developed nested triangular patterns. Based on water condensation studies on the nested triangular patterns, they reported that the nested pattern increases the droplet speed.

In this chapter, an overview of systematic studies of water harvesting from condensation of water vapor and/or fog, using a triangular geometry carried out by Song and Bhushan (2019a, b, c, d), Feng and Bhushan (2020) and Bhushan and Feng (2020), is presented.

5.1   Experimental Details

The triangular patterns of various geometry with hydrophilic surfaces have been used. Surrounding regions with less wettability than the triangular patterns were selected to constrain collected water in the pattern. For water condensation studies, a low temperature of 5 °C was used to decrease saturated vapor pressure to promote water condensation. For fog collection studies, a commercial humidifier was used to generate a stream of fog.

5.1.1   Fabrication of Water Collection Surfaces

5.1.1.1   Single Triangular Patterns and Triangular Array

Two types of samples were fabricated. One type of sample contained a single triangular pattern, which was used to investigate the movement of the droplets, shown in Fig. 5.1a (Song and Bhushan 2019a, b). The single triangular pattern with a 20 mm length, with region A being hydrophilic, was surrounded by a superhydrophobic rim (0.5 mm wide). The superhydrophobic rim was produced to serve as a dam to the condensed liquid in the triangular pattern. If any droplets are transferred in the superhydrophobic dam region, they will slide to the outer less wettable region and will be prevented from sliding back into the pattern. Three included angles ($\alpha = 5°$, $9°$ and $17°$) were selected to investigate the effect of included angle on the droplet condensation and transport process. A rectangular reservoir was provided at the end of pattern in order to study the time required for droplet transport to the reservoir. The other type of sample contained an array of triangular patterns that were located on both sides of a rectangular reservoir, to increase the amount of the collected water for better measurement accuracy, shown in Fig. 5.1b. An array with an included angle of 9° and length of 10 mm, was used. To study the effect of included angle, arrays with four included angles of 9°, 17°, 22° and 30°, with a length of 10 mm were selected. To study the effect of length, arrays with four lengths of 5, 10, 20 and 30 mm with an included angle of 9° were selected.

The triangular patterns were fabricated on a hydrophilic glass slide (Song and Bhushan 2019a, b). To fabricate the patterned sample, the boundaries of pattern

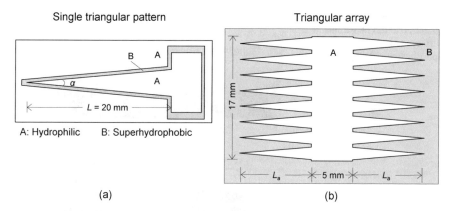

Fig. 5.1 Schematics of a single triangular pattern, and an array of triangular patterns with a rectangular reservoir (adapted from Song and Bhushan 2019b)

B were printed on a paper that was placed under the glass slide and a piece of adhesive tape was put on top of the glass slide. Next, region B was cut onto the adhesive tape, guided by the pattern on the paper underneath, so that region B was exposed to air and region A was protected by the tape. Then, a superhydrophobic coating was spray coated on the glass slide, followed by removal of the adhesive tape that covered region A. The superhydrophobic coating consisted of 10 nm hydrophobic SiO_2 nanoparticles (Aerosil RX300) and a binder of methylphenyl silicone resin (SR355S, Momentive Performance Materials), both of which were pre-mixed in acetone before spraying (Bhushan 2018; Bhushan and Martin 2018). Region B after coating became superhydrophobic and region A remained hydrophilic.

To compare water collection on a triangular pattern with that of rectangular pattern, glass slides with hydrophobic, triangular and rectangular patterns surrounded by superhydrophobic surfaces were also produced (Song and Bhushan 2019a). To produce a hydrophobic pattern, a monolayer of perfluorodecyl-trichlorosilane (FTDS) (448931, Sigma-Aldrich) was deposited on the glass surface using a vapor deposition method (Bhushan 2018).

To study the effect of various wettabilities, three wettabilities were studied. These included hydrophobic and hydrophilic patterns surrounded by superhydrophobic regions and superhydrophilic patterns surrounded by hydrophobic regions. The fabrication process for the first two patterns has already been described. To fabricate a superhydrophilic pattern surrounded by a hydrophobic region, a hydrophobic glass slide was used. The surrounding region was masked by placing tape on it to preserve the hydrophobic glass surface. Then to make the triangular pattern superhydrophilic, it was exposed to ultraviolet-ozone (UVO) treatment for 1 h (Song and Bhushan 2019a).

The wettability of various surfaces was measured using an automated goniometer (Model 290, Ramé-hart Instrument Co.). The wettability characterization data for

Fig. 5.2 Wettability
characterization of various
surfaces (adapted from Song
and Bhushan 2019a)

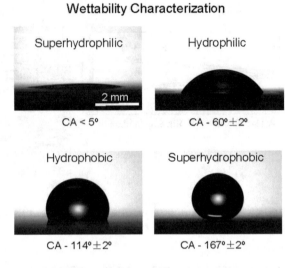

various surfaces is presented in Fig. 5.2 (Song and Bhushan 2019a). The measurement technique of static contact angle (CA) has been described in Chap. 4.

5.1.1.2   Rectangular and Triangular Patterns with Multistep Wettability Gradient

Samples with rectangular and triangular patterns and multistep wettability gradient were fabricated on polydimethylsiloxane (PDMS) coated glass slides (Feng and Bhushan 2020). For deposition of PDMS (Sylgard® 184, Dow Corning) coating, a silicone elastomer curing agent was mixed with the base at the weight ratio of 1:10. The uncured PDMS was degassed for 15 min at room temperature to remove entrained air bubbles. The mixture was subsequently poured onto a glass slide and allowed to flow for 1 h to form approximately 3 mm thick flat samples. The mixture was cured by heating at 80 °C for 2 h. Finally, the desired samples were produced by cutting the cured PDMS.

PDMS surfaces are hydrophobic with a contact angle of about 110°. UVO treatment is commonly used to activate surfaces to change their degree of wettability (Bhushan 2018). The wettability gradient with various degrees of wettability can be achieved by treating various segments of a pattern for different lengths of treatment times. The UVO exposure was generated from a U-shaped, ozone-producing and ultraviolet lamp (18.4 W, Model G18T5VH-U, Atlantic Ultraviolet Co.) and samples were placed 100 mm underneath the light source. By using a moving mask in selected steps, samples were treated for various times, to produce surfaces with wettability gradient (Fig. 5.3a) (Feng and Bhushan 2020). The mask was moved every 10 min with the first step of 2.5 or 5 mm and the other five steps of 1.5 or 3 mm.

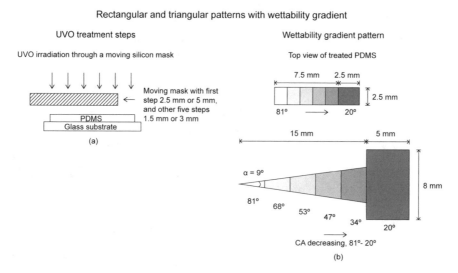

Fig. 5.3 Schematics (**a**) of fabrication steps of the PDMS surfaces with multistep wettability gradient by UVO treatment, and **b** rectangular and triangular patterns with wettability gradient (adapted from Feng and Bhushan 2020)

Figure 5.3b shows the schematics of rectangular and triangular samples with wettability gradients (Feng and Bhushan 2020). The size of the rectangular surface was 10 mm × 2.5 mm. The triangular surface was 20 mm long and 8 mm wide at the base. The rectangular base was 5 mm long and the triangular region was 15 mm long with an included angle of 9°. Both rectangular and triangular samples had 6 wettability steps. The storage steps were 2.5 or 5 mm long, and the other five steps were 1.5 mm or 3 mm long on rectangular and triangular samples, respectively. From left to right, the CA of the surfaces decreased in six steps (81°, 68°, 53°, 47°, 34° and 20°).

5.1.1.3 Nested Triangular Patterns

The nested triangular patterns were formed by nesting two triangles with an included angle of 17° and length of 10 mm each, as shown in Fig. 5.4 (Bhushan and Feng 2020). The initial width of the second triangle at the junction at a length of 10 mm was ¾ width of the first triangle. To minimize resistance to droplet movement at the junction, entrance angle to the second triangle was set to be about 5°. As will be discussed later, a configuration of nested triangles with an initial width of the second triangle at the junction, with ½ width of the first triangle, was also tried, but that impeded droplet movement. The total length of the entire nested pattern was 20 mm. For comparison, the pattern with a single triangle was used with an included angle of 17° at a length of 20 mm (Fig. 5.4).

Fig. 5.4 Schematics of patterns with a single triangle, and with two nested triangles (adapted from Bhushan and Feng 2020)

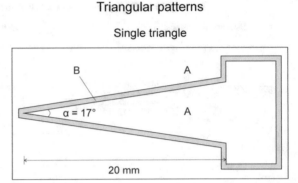

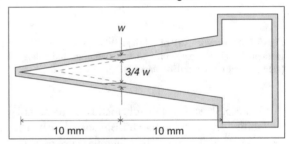

5.1.2 Experimental Apparatuses for Water Collection

5.1.2.1 Water Collection from Condensation

Figure 5.5 shows a schematic of the apparatus for water collection from condensation (Song and Bhushan 2019a, b). The samples were placed on top of an aluminum block on the horizontal table and cooled by a thermoelectric Peltier cooler (GeekTeches) down to about 5 ± 1 °C. A digital camera (Koolertron, 5MP 20-300X) was used to record the water collection process.

A property of interest in water condensation is dew point temperature. It is the temperature to which air must be cooled to become saturated with water vapor. When further cooled below the dew point, airborne water vapor will condense on the surface to form water droplets (dew). At a temperature lower than dew point the saturated vapor pressure of water in the ambient air decreases which leads to water condensation (Alduchov and Eskridge 1996). The dew point temperature, T_d, can be calculated using the following equation (Lawrence 2005; Pruppacher and Klett 2010; Moran et al. 2018).

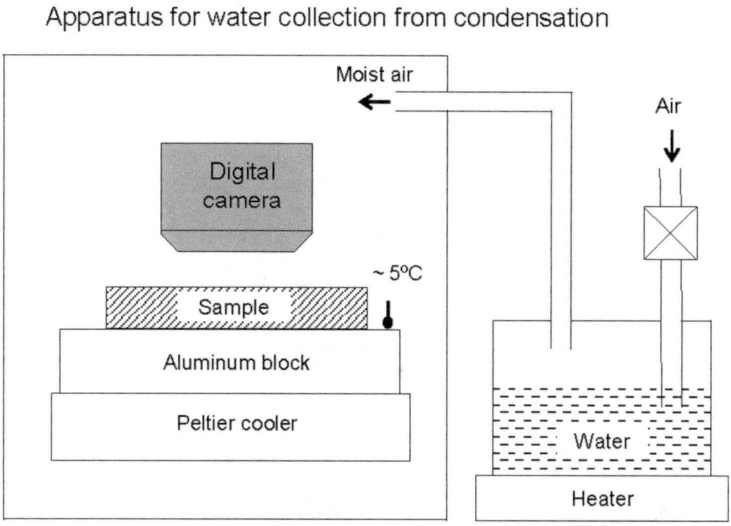

Apparatus for water collection from condensation

Fig. 5.5 Schematic of apparatus for water collection from condensation (adapted from Song and Bhushan 2019a, b)

$$T_d = T - ((100 - \text{RH})/5) \qquad (5.1)$$

where RH is relative humidity of air-water mixture defined as partial pressure of water vapor in the mixture divided by saturation vapor pressure. Using (5.1), the dew point temperature at the ambient (22 °C and 50% RH) is about 12 °C, higher than the sample temperature of 5 °C.

The samples were housed in an acrylic chamber (1 m × 0.5 m × 0.8 m). The air temperature in the chamber was 22 ± 1 °C. Relative humidity (RH) was controlled by injection of humid air. The humid air was produced by an air stream that passed through a tank of hot water. By changing the temperature of the hot water and the flow rate of the air stream, RH in the chamber could be controlled between 30 and 95%. A digital camera (Koolertron, 5MP 20-300X) was used to capture the condensation process.

Experiments were conducted at a relative humidity of 85 ± 5%, except those performed to study the effect of relative humidity. When the sample was cooled down to 5 °C at RH >50%, which is lower than the dew point of the ambient air, water vapor continuously condensed on the triangular region.

5.1.2.2   Water Collection from Fog

Figure 5.6 shows a schematic of the apparatus for water collection from fog (Song and Bhushan 2019c). A commercial humidifier (Crane, EE-3186) was used to generate a

Apparatus for water collection from fog

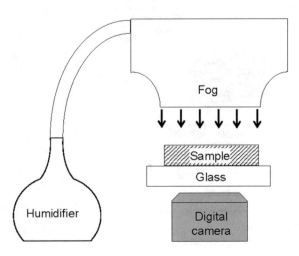

Fig. 5.6 Schematic of apparatus for water collection from fog (adapted from Song and Bhushan 2019c)

stream of fog which was injected into a box. A rectangular opening at the bottom of the box shaped the fog flow into a rectangular channel about 40 mm × 25 mm. The samples were placed on top of a piece of transparent glass (3 mm × 150 mm 150 mm). Experiments were conducted in ambient conditions with a temperature of 22 ± 1 °C and relative humidity (RH) between 35 and 50%. A digital camera (Koolertron, 5MP 20-300X) was used to record the water collection process.

5.1.2.3 Water Collection from Fog and Condensation

Figure 5.7 shows a schematic of the apparatus to collect microdroplets from the fog as well as nanodroplets of condensed water vapor (Song and Bhushan 2019d). The samples were placed on top of an aluminum block that was cooled by a Peltier cooler down to about 5 ± 1 °C. The fog was generated in the box by a commercial humidifier (Crane, EE-3186). Experiments were conducted in ambient with a temperature of 22 ± 1 °C. Because of the fog, relative humidity was expected to be close to 100%.

Apparatus for water collection from fog and condensation

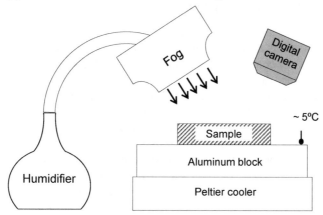

Fig. 5.7 Schematic of apparatus for water collection from fog and condensation (adapted from Song and Bhushan 2019d)

5.1.2.4 Water Collection Measurements

To measure the mass of the collected water for both types of samples, after the test, a piece of tissue paper was used to absorb the collected water. The paper before and after soaking was weighed by a microbalance (Denver Instrument Company No. B044038). The microbalance could measure a minimum mass of 1 mg. The mass of tissue paper piece ranged from about 100–150 mg. The mass of the collected water was about 100–200 mg. The mass of collected water could be measured with an accuracy of about ±5%.

5.2 Results and Discussion—Water Condensation Studies

To study the role of a triangular pattern on droplets' mobility, experiments were conducted on triangular and rectangular patterns. To study the effect of wettability, experiments were conducted on triangular patterns with different wettabilities. To understand the droplet transport mechanism, single droplets of different volumes were deposited on a triangular pattern at room temperature, and the droplet movements were observed. Next, to study the effect of geometry of triangular patterns and relative humidity on the water collection rate from condensation, experiments were conducted using triangular patterns with different geometries at a range of relative humidity. For measurement of water collection rates, a reservoir surrounded by an array of triangular patterns was used to collect a larger amount of water.

5.2.1 Rectangular Versus Triangular Pattern and Various Wettabilities

To compare the effect of rectangular and triangular patterns on droplet condensation and mobility, water condensation studies on hydrophobic rectangular and triangular patterns surrounded by superhydrophobic surfaces were carried out. The data is shown in Fig. 5.8 (Song and Bhushan 2019a). Nucleation occurs with the initial radius of the droplets on the nanoscale and these coalesce and grow to micro-droplets with time (Sigsbee 1969). In the rectangular pattern, the micro-sized droplets formed rather uniformly in the hydrophobic area. With time, the micro-droplets grew and coalesced resulting in larger droplets (numbered 1–6). When the droplets grew big enough to touch each other, they coalesced into even larger ones (1 + 2, 3 + 4 and 5 + 6). However, the center position of the merged droplets did not change, because there was no transportation after the coalescence.

In the triangular pattern, the initial formation of the micro-sized droplets was similar to that on a rectangular pattern. Tiny droplets (1, 2 and 3) shown in Fig. 5.8, coalesced to a larger one (1 + 2 + 3), which is larger than the local width of the triangular pattern. The merged droplet moved in the direction of the wider triangular width (Song and Bhushan 2019a). Condensation and coalescence continued. The newly formed droplets kept on growing and a new cycle of coalescence made the droplet move even farther.

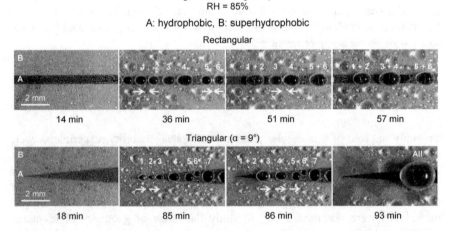

Fig. 5.8 Selected optical images of condensed droplets on hydrophobic rectangular and triangular patterns surrounded by superhydrophobic coatings. On the rectangular pattern, the condensed droplets do not move, but grow and coalesce with a size larger than the pattern. However, the droplets on the triangular pattern move in the direction with larger triangular width when tiny droplets coalesce into a droplet with a size larger than the local triangular width. Arrows shown below some droplets are based on the movement of center of the droplet observed in videos (adapted from Song and Bhushan 2019a)

To study the effect of wettability, the triangular regions with three wettabilities and surrounding regions with less wettability were studied. Data for hydrophobic, hydrophilic and superhydrophilic patterns are shown in Fig. 5.9a (Song and Bhushan 2019a). Wettability of the triangular patterns was found to affect the condensation process and droplet mobility. In the case of the hydrophobic pattern, droplets were more spherical and did not spread. However, they touched the pattern boundaries readily, which is necessary for droplet movement. In the case of superhydrophilic and hydrophilic patterns, droplets spread and touched the pattern boundaries to facilitate droplet movement. However, in the case of superhydrophilic patterns, condensed water spread in the form of a thin film, making the surface transparent (showing the black background of substrate underneath). Part of the water film showed a white region due to the light reflection at the right side, as marked. The thin film evaporates readily, and is not desirable for water collection. Furthermore, adhesion of the water droplet is high which impedes droplet movement.

The time it takes to initiate self-transport is affected by the wettability as well. On the hydrophilic triangular pattern, it takes about 25 min for the self-transportation of condensed droplets to occur from the tip to a distance of about 5.5 mm, whereas it takes about 94 min on the hydrophobic pattern to move the same distance. Therefore, hydrophilic triangular patterns surrounded by superhydrophobic regions are most efficient in self-transport of droplets with three types of wettability (Song and Bhushan 2019a).

To clearly illustrate water condensation and transport process as a function of time, Fig. 5.9b presents schematics of droplet nucleation and movement (Song and Bhushan 2019a). During condensation, the droplets nucleate and coalesce. Once they touch the two sides, they start to move because of Laplace pressure gradient.

All following data is presented for hydrophilic patterns.

5.2.2   Single Droplet Experiments on Hydrophilic Triangular Patterns

To understand the droplet transport mechanism, single droplet experiments were conducted by depositing a droplet using a pipette at the inside of the tip of the triangular pattern with varying volumes from 10 to 100 μL. Increments of 5 μL were used for volumes ranging between 10 and 50 μL and 10 μL ranging between 50 and 100 μL. After deposition, the droplet moved along the pattern due to Laplace pressure gradient, and stopped after traveling some distance. After the first droplet had stopped, the pipette was used to add additional droplets at the current location of the previous droplet, not at the tip of the pattern. Increments were added until a maximum value of 100 μL had been added.

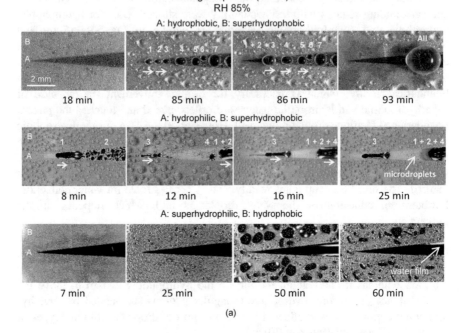

Triangular pattern (α = 9°)
RH 85%
A: hydrophobic, B: superhydrophobic

| 18 min | 85 min | 86 min | 93 min |

A: hydrophilic, B: superhydrophobic

| 8 min | 12 min | 16 min | 25 min |

A: superhydrophilic, B: hydrophobic

| 7 min | 25 min | 50 min | 60 min |

(a)

Schematic showing droplet nucleation and movement
A: hydrophilic, B: superhydrophobic

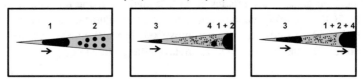

Droplets touch boundaries and move because of Laplace pressure gradient

(b)

Fig. 5.9 a Selected optical images of condensed droplets on three triangular patterns with different wettability. The condensed droplets on the triangular patterns move in the direction with larger triangular width when tiny droplets coalesce into one droplet with a size larger than the local triangular width. Arrows shown below some droplets are based on the movement of the center of the droplet observed in the videos. On the hydrophobic triangular pattern, the constrained droplets are closer to being spherical compared to the ones on the hydrophilic triangular pattern where the droplets spread into long stripes and travel faster than those on the hydrophobic surface. On the superhydrophilic triangular pattern surrounded by the hydrophobic coating, condensed water droplets spread into a thin film during the entire condensation process. Hydrophilic triangular patterns surrounded by the superhydrophobic regions are most effective in self-transport. **b** Schematic illustrations of droplet nucleation and movement on a hydrophilic triangular pattern surrounded by a superhydrophobic region (adapted from Song and Bhushan 2019a)

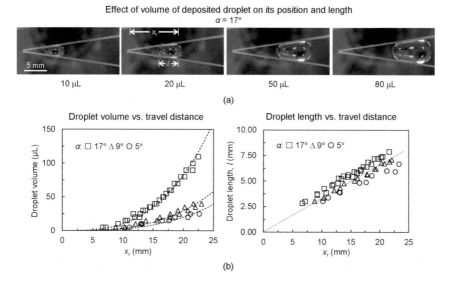

Fig. 5.10 **a** Selected optical images of the deposited droplets with different volumes transported along the triangular pattern when they stop. **b** Droplet volume and length as a function of travel distance. These data were taken when the droplet stopped (adapted from Song and Bhushan 2019b)

Figure 5.10a shows the optical images of the locations of the stopped droplets of different volumes for a triangular pattern with an included angle of 17° (Song and Bhushan 2019b). A droplet with larger volume traveled farther.

To mathematically understand the role of Laplace pressure gradient on the droplet movement and the distance traveled, a droplet placed on the hydrophilic triangular pattern, as shown in Fig. 5.11 was analyzed (Song and Bhushan 2019b). In the schematic, $w(x)$ is the local width of the triangle at a distance x from the tip of the triangle. The droplet was constrained by the superhydrophobic region and became wedged shaped. The local radius of the curvature of the droplet along the triangle can be written as

Droplet constrained within a triangular pattern

Fig. 5.11 An optical image and schematic of a droplet constrained within the triangular pattern (adapted from Song and Bhushan 2019b)

$$R(x) \sim w(x)/(2\sin\,\theta(x)) \tag{5.2}$$

where $\theta(x)$ is the contact angle at the boundaries. The Laplace pressure generated by the local curvature is given as

$$\Delta P = \gamma_{LA}/R(x) \sim 2\gamma_{LA}\sin\,\theta(x)/w(x) \tag{5.3}$$

where γ_{LA} is the surface tension of water in air (Adamson and Gast 1997; Isrealachvili 2011). For the constrained droplet, $w(x)$ increases from the narrower side to the wider side, and hence the Laplace pressure at the narrower side is larger than the wider side. As a result, a driving force is generated to transport the droplet with the direction pointing to the wider side. The driving force of the Laplace pressure exists as long as the droplet is large enough to contact both boundaries of the triangular pattern. When the droplet moved farther in the triangular pattern, the magnitude of the driving force decreases because of the decrease in the gradient of the inverse of local width of the triangle. The droplet stops when the driving force is smaller than the adhesion force.

Figure 5.10b shows a plot of the droplet volume as a function of travel distance, measured from the tip of the triangle to the right hand edge of the droplet (x_r) at various included angles (Song and Bhushan 2019b). The distance was measured to the right hand edge because the droplet will be sucked to the reservoir once the right edge touches the reservoir. The distance was measured when the droplet stopped. A droplet with a larger volume traveled farther. A droplet with a given volume was transported farther by a triangle with a smaller included angle. For example, to move a droplet 20 mm to the right hand edge, the droplet volume had to reach 82, 29 and 20 μL on the triangular pattern with the included angle $\alpha = 17$, 9 and 5°, respectively.

Figure 5.10b also shows the relationship between the length of the stopped droplet (l) and its position (x_r) (Song and Bhushan 2019b). The data shows that l increased linearly with x_r and the included angle had little effect. It was observed that the droplets were elongated along the travel direction. The droplet elongation was believed to occur due to the adhesive force, which is directly related to the contact angle hysteresis (difference between advancing and receding contact angles) in the triangular area.

5.2.3   Hydrophilic Triangular Patterns—Effect of Geometry and Relative Humidity

Water condensation, coalescence, and transport process was investigated for hydrophilic triangular patterns with various geometry and at various relative humidity.

5.2.3.1 Included Angles

Figure 5.12 shows the condensed droplets on triangular patterns with three included angles (Song and Bhushan 2019b). In the beginning of condensation, the condensed droplets were relatively small, as shown in the first column of Fig. 5.12a. As the condensation continued, the growing droplets started to coalesce into bigger droplets. Eventually, they were big enough to touch the superhydrophobic borders, which triggered the motion driven by the Laplace pressure gradient, as shown in the second and third column of the figure. For example, at a time of 45 min after the start of condensation on the triangular pattern with $\alpha = 9°$, there were five millimeter-sized droplets (numbered 1–5) that touched the borders. At 76 min, droplets 1 and 2 coalesced into one big droplet $(1 + 2)$ and droplets 3 and 4 coalesced into another $(3 + 4)$. After the coalescence of droplets 1 and 2, the droplet volume increased and the Laplace pressure gradient was able to drive the droplet. Due to adhesion, the droplet $(1 + 2)$ was elongated and stopped after moving a step of 1.4 mm based on the calculation of the center of the droplet area. It is further noted that the position of the center of droplets 3 and 4 did not change after their coalescence. This is because the size of the coalesced droplet $(3 + 4)$ was too small and the Laplace pressure gradient along the coalesced droplet could not overcome the adhesion.

The length of the droplet as a function of travel distance x_r after it stops, is shown in Fig. 5.12b (Song and Bhushan 2019b). Similar to Fig. 5.10b, the length of the elongated droplet increases linearly with its position x_r, and the included angle has little effect on the length.

To study the condensation rate, one needs to know the size (mass or volume) of the droplet and time taken to reach the reservoir. Figure 5.12c shows the droplet mass and time needed for the droplet to reach the reservoir through the entire triangular pattern with different included angles (Song and Bhushan 2019b). As α increases, it takes more time for the droplet to be transported to the reservoir. However, for larger α, the mass of a coalesced droplet which starts to move is larger, as shown in the figure.

To study transport efficiency of the condensed droplet across the triangular pattern, an array of triangular patterns was used to increase the amount of collected water for high accuracy. This will be presented in a later section.

5.2.3.2 Relative Humidity

Figure 5.13a shows the condensed droplets at two different values of relative humidity on a triangular pattern with an included angle of 9° (Song and Bhushan 2019a, b).The droplets coalesced as a function of time and once they touched the borders, they started to move. The coalesced droplets eventually moved to the reservoir and the sizes of the droplets were weighed by soaking a slice of paper tissue and weighing it. Figure 5.13b shows the mass of the final droplet right before reaching the reservoir. RH did not affect the mass of the final droplet before

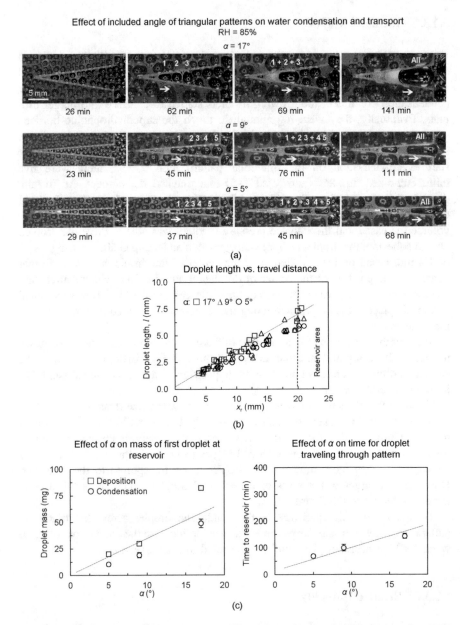

Fig. 5.12 a Selected optical images of the condensed droplets on a single triangular pattern with three included angles at different times. Arrows shown below some droplets are based on movement of center of the droplet observed in videos. **b** Length of the coalesced droplet (*l*) as a function of the travel distance, x_r, after it stops. **c** Effect of included angle on the mass of the first droplet at the reservoir and the time taken for the droplet traveling through the pattern (adapted from Song and Bhushan 2019b)

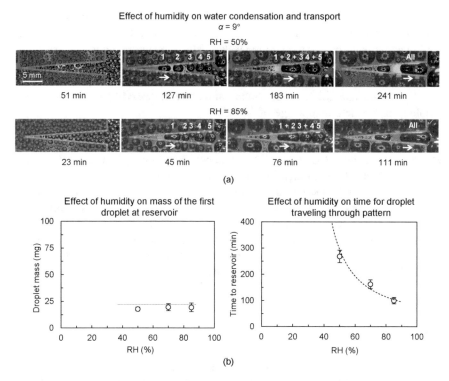

Fig. 5.13 a Selected optical images of the condensed droplets on a single triangular pattern in different relative humidity at different times. Arrows shown below some droplets are based on movement of center of the droplet observed in videos. **b** Effect of relative humidity on the mass of the first droplet in the reservoir and the time taken for the droplet traveling through the pattern (adapted from Song and Bhushan 2019b)

reaching the reservoir. Figure 5.13b also shows the time it takes for the droplet to reach the reservoir at different values of relative humidity. The travel time through the reservoir decreased with an increase in RH because of increased condensation (Song and Bhushan 2019b).

5.2.4   Array of Hydrophilic Triangular Patterns

To increase the water collection rate, samples containing a reservoir with an array of triangular patterns were used. Figure 5.14a shows the optical image of the reservoir with triangular patterns with 10 mm length and an included angle of 9°, after condensation for 450 min (Song and Bhushan 2019b). The condensed water on the triangular area was transported to the reservoir, and its mass was measured using a paper tissue. To evaluate the additional mass of water condensed at the reservoir area, a rectangular hydrophilic area was placed beside the array and the mass of the condensed water was measured as well. When calculating the condensation rate on the triangular patterns, the mass of the water on the rectangular region was deducted from the mass of the water on the reservoir with array.

The effect of relative humidity on the condensation rate is shown in Fig. 5.14b (Song and Bhushan 2019b). The condensation rate increased linearly with relative humidity in the measured range of 50–85%.

Next, for optimization of water collection designs, the effect of the included angle, α, and the length of the triangular patterns, L_a, were studied. As shown in Fig. 5.14c, the included angle did not affect the condensation rate (Song and Bhushan 2019b). Even though the droplet transported slowly on a triangle with a larger included angle, the size of the droplet was larger which may provide similar condensation rates. Figure 5.14c also shows the condensation rate as a function of the length of the triangular patterns. The condensation rate decreased when the length increased. Since a shorter distance requires less time to transport the condensed droplets and though droplets being removed are expected to be smaller, the removal rate increased the condensation rate.

5.2.5 Summary

The water condensation and transport process of triangular patterns was investigated. The effect of different wettability of triangular patterns surrounded by the region with less wettability was studied. The hydrophilic triangular patterns surrounded by superhydrophobic regions were found to be desirable for water collection. For droplet transport, when the droplets were constrained within the triangular patterns, they started to move after coalescence, and they reached a critical size and touched the two sides. The triangular pattern with a larger included angle needed more time to transport condensed droplets, however, the mass was larger. A water collection reservoir was fabricated with an array of triangular patterns to measure the condensation rate. The relative humidity increased the condensation rate. The included angle did not affect the condensation rate. A decrease in the length of the patterns increased the condensation rate (Song and Bhushan 2019a, b).

To design a condensation water collection tower, a hydrophilic pattern surrounded by a superhydrophobic region should be used. Since the included angle had no effect on condensation rate and a shorter length promoted condensation, a larger number of triangular patterns with smaller included angles and shorter lengths can be used for a higher condensation rate. Relative humidity of the ambient air increased the condensation rate, therefore, high relative humidity is desirable. However, it may not be possible to control it in practical applications (Song and Bhushan 2019a, b).

Effect of humidity, included angle and length of triangular
patterns on water condensation rate

Reservoir with 16 triangular patterns, $\alpha = 9°$, $L_a = 10$ mm
RH = 50%, at 450 min

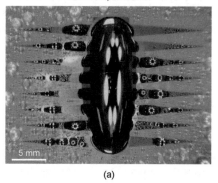

(a)

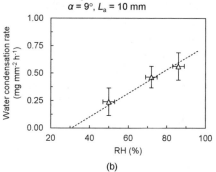

(b)

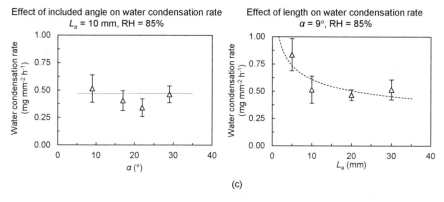

(c)

Fig. 5.14 **a** Representative optical image of the condensed water in the reservoir surrounded by an array of triangular patterns, **b** Water condensation rate as a function of relative humidity, and **c** Water condensation rate as a function of included angle and length of the triangular patterns (adapted from Song and Bhushan 2019b)

5.3   Results and Discussion—Fog Water Collection Studies

To study the effect of geometry of triangular patterns in fog, the water collection studies were conducted on hydrophilic triangular patterns for various included angles. For measurement of water collection rates, a reservoir surrounded by an array of triangular patterns was used to collect a larger amount of water.

5.3.1   Included Angles

Figure 5.15a shows the water collection and transport process on triangular patterns with different included angles (Song and Bhushan 2019c). The included angle affects the water transport process. The time taken to transport the water droplets across the triangular pattern increases with an increase in the included angle. For example, to transport the collected water to reach the reservoir, $x_r = 20$ mm, it takes about 78, 57 and 42 min on the patterns with included angles of 17°, 9° and 5°, respectively. The size of droplet needed to trigger the transport of the deposited droplet, increases with an increase in the included angle.

Due to adhesion of the hydrophilic pattern to the droplet, the collected droplet was elongated before it could start to move. The length of the elongated droplet (l) right after the droplet stopped moving as a function of the travel distance, x_r, is shown in Fig. 5.15b (Song and Bhushan 2019c). Length increased linearly with x_r, and the included angle had no effect. These results agree with observations made in the water condensation study presented in the previous section.

Figure 5.15c shows the droplet mass and time needed for the droplet to reach the reservoir through the whole triangular pattern as a function of included angles (α) (Song and Bhushan 2019c). As α increases, the mass of the droplet that reached the reservoir increased, while it took more time for the droplets to reach the reservoir. These results also agree with observations made in the water condensation study reported earlier.

5.3.2   Array of Triangular Patterns

To increase the water collection rate, samples containing a reservoir with an array of patterns were used. Figure 5.16a shows an optical image of the reservoir with an array of triangular patterns with 10 mm length and an included angle of 9°, after being exposed to fog for one hour (Song and Bhushan 2019c). The collection rate of water was about 0.86 mg mm^{-2} h^{-1}.

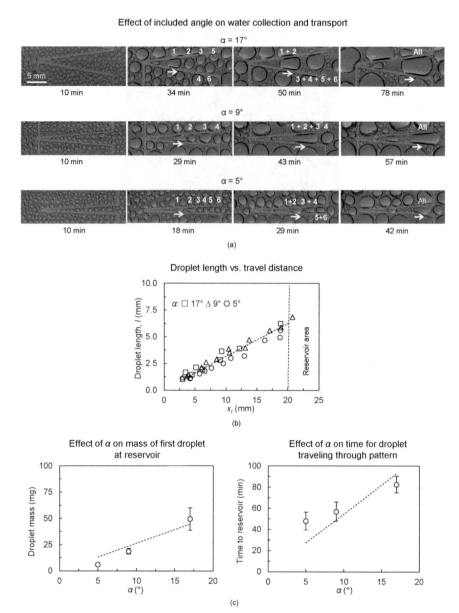

Fig. 5.15 a Selected optical images of droplets on single triangular patterns with different included angles at different times under fog. Arrows shown below some droplets correspond to movement of center of the droplet observed in videos. **b** Length of the coalesced droplet (l) as a function of the travel distance (x_r) after it stops. **c** Effect of included angle (α) on the mass of the first droplet at the reservoir and time taken for the droplet traveling through the pattern (adapted from Song and Bhushan 2019c)

Effect of included angle and length of patterns on water collection rate

Reservoir with 16 triangular patterns
$\alpha = 9°$, $L_a = 10$ mm, 60 min

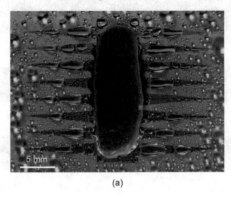

(a)

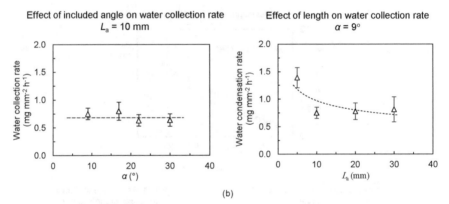

(b)

Fig. 5.16 a Representative optical image of the water collection in the reservoir surrounded by an array of triangular patterns in fog, and **b** water collection rate as a function of included angle and length of the triangular patterns (adapted from Song and Bhushan 2019c)

For optimization of water collector design, the effect of the included angle, α, and the length of the triangular patterns, L_a, were studied. Figure 5.16b shows that the included angle did not affect the water collection rate. Even though the droplet transported slowly on a triangle with a larger included angle, the size of the droplet was larger which may provide the similar collection rates. Figure 5.15b also shows the water collection rate as a function of the length of the triangular patterns (Song and Bhushan 2019c). The water collection rate decreased when the length increased. Since a shorter distance requires less time to transport the droplets and although droplets being removed are small, the removal rate increases the collection rate.

5.3.3 Summary

The water collection rates and transport ability of the triangular patterns from fog were investigated. Hydrophilic triangular patterns were surrounded by a rim of superhydrophobic regions. When exposed to fog, droplets accumulate on hydrophilic patterns. Droplets grow and start to coalesce into bigger ones. Eventually, they are big enough to touch the superhydrophobic borders, which triggers the transport motion. A wedge-shaped droplet generates a Laplace pressure gradient that is able to spontaneously drive the droplet. The collected water moves slower on a triangular pattern with a larger included angle, however, larger water droplet is transported to the reservoir. In experiments with an array of triangular patterns surrounding a reservoir, the water collection rates were measured. Included angle had no effect on the collection rate, however it increased with a decrease in the length of the pattern.

5.4 Results and Discussion—Fog Water Collection and Condensation Studies

The droplet collection process from fog deposition and/or vapor condensation was investigated first on a flat hydrophilic surfaces. Then the samples with triangular patterns were studied to measure the water collection rate from the fog and/or condensation.

5.4.1 Flat Hydrophilic Surfaces Under Different Conditions

Figure 5.17a shows the droplet collection process on hydrophilic surfaces under different conditions (Song and Bhushan 2019d). Under fog, the microdroplets suspended in the fog gets deposited on the hydrophilic surface. The droplets become larger due to the coalescence of the microdroplets. Under the condensation condition, the droplets nucleate on the surface. The condensation process of water vapor is driven by the phase change which starts with nanodroplets (Pruppacher and Klett 2010). The condensation process is dependent on the degree of subcooling and the relative humidity. Again the droplets coalesce and become larger.

In the experimental apparatus used, the droplet size in fog appears to be larger than in the condensation where the substrate temperature is 5 °C with a relative humidity of 85 ± 5% or more.

When the condensation was included in the fog flow, the size of the water droplets appears to have increased, as shown in Fig. 5.17b. It took around 5 min for the droplet to reach the size of around 1 mm while it took about 15 min and 25 min under fog and condensation conditions, respectively, for the droplets to reach a similar size, as shown in Fig. 5.17b (Song and Bhushan 2019d).

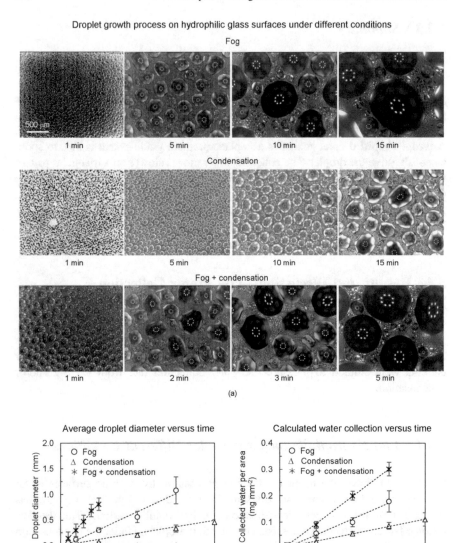

Fig. 5.17 a Selected optical images of droplets on flat hydrophilic surfaces, under fog, condensation, and fog + condensation, b average diameter of the coalesced droplets at different times, and c the mass of collected water per unit area at different times. The mass in (c) was calculated from the images in (a) and droplets were assumed to be spherical (adapted from Song and Bhushan 2019d)

The mass of the collected water on the hydrophilic surface was estimated by assuming that the droplets were spherical (Song and Bhushan 2019d). The mass is given as:

$$\rho \sum_{i}^{N} \frac{\pi(2 + \cos\theta)(1 - \cos\theta)^2}{24\sin^3\theta} d_i^3 \qquad (5.4)$$

where N is the total number of the coalesced droplets per unit area, d_i is the diameter of the ith droplet observed by the top view image, $\theta = 61°$ is the contact angle and $\rho = 998$ kg/m^3 is the water density (Rumble 2019). The estimated water collection per unit area are shown in Fig. 5.17c (Song and Bhushan 2019d). With both condensation and fog, the water collection rate was increased by more than two times the water collection rate in fog alone.

The water collection process under different conditions is shown schematically in Fig. 5.18 (Song and Bhushan 2019d). Under the fog condition, microdroplets

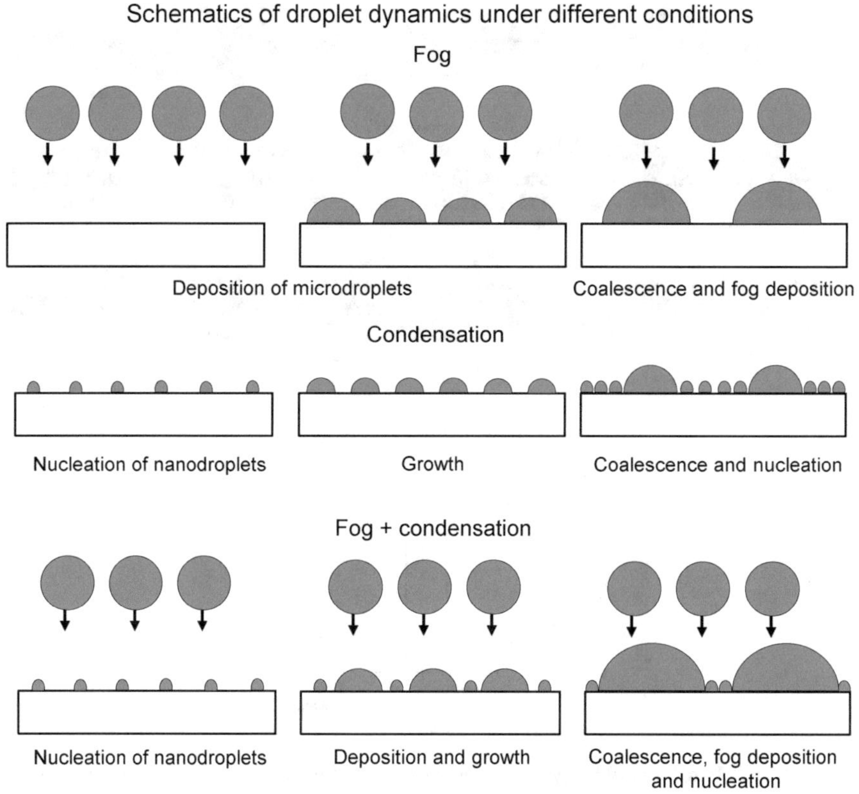

Fig. 5.18 Schematic illustration of proposed droplets growth process under fog, condensation, and fog + condensation (adapted from Song and Bhushan 2019d)

carried by the fog flow are deposited on the substrate without any phase change. The droplet growth is dominated by the feed of microdroplets and coalescence. Under the condensation condition, nucleation occurs on the substrate where the initial radius of the nucleated droplet is on nanoscale (Sigsbee 1969). When combining fog and condensation, both microdroplet deposition and nanoscale nucleation exist on the substrate. The microdroplets from fog and nanodroplets from condensation coalesce more rapidly resulting into larger droplets.

5.4.2 Triangular Patterns Under Different Conditions

Figure 5.19 shows the collection and transport of water droplets moving along the triangular pattern with an included angle of 9° under different conditions (Song and

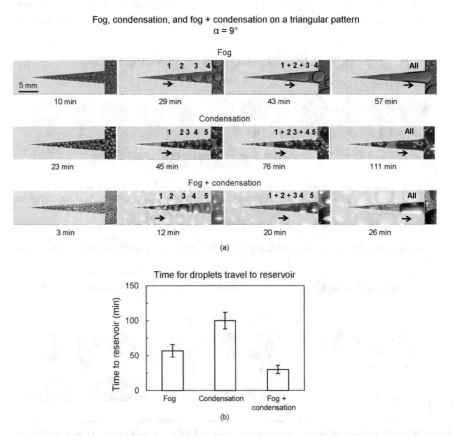

Fig. 5.19 a Selected optical images of the droplets on triangular patterns at different times, under fog, condensation, and fog + condensation. Arrows shown below some droplets represent movement of center of the droplet observed in videos. **b** Time taken for the droplet traveling through the pattern to the reservoir (adapted from Song and Bhushan 2019d)

Bhushan 2019d). It has been reported earlier that the hydrophilic triangular pattern can drive the droplet by the Laplace pressure gradient without an external force. The mechanism of the driving force of the droplet is similar under different conditions. In the beginning, small droplets grew and coalesced at the tip area of the triangle, until there was one droplet large enough to touch the superhydrophobic borders. A large droplet was elongated in the triangular direction and the vertical width varied, which triggered the Laplace pressure gradient and drove the droplet rightward.

The key to transport of the droplet at a fast speed is to promote the droplets to grow large enough to touch the triangular borders as soon as possible. Under the condition of fog or condensation alone, the droplets grew at a lower speed than the ones with both fog and condensation as mentioned in the previous section. As a result, it took only 26 min for the droplet to move across the entire pattern under fog and condensation while it took 57 and 111 min for the droplet under fog or condensation, respectively.

5.4.3 Array of Triangular Patterns Under Fog and Condensation

To study the water collection rates of the triangular patterns under the condition of fog and condensation, an array of triangular patterns was fabricated to surround a reservoir, so that the collected water can be increased. Figure 5.20a shows a representative optical image of the reservoir with an array of triangular patterns with 10 mm length and included angle of 9° after being exposed to fog and condensation for 18 min (Song and Bhushan 2019d).

Figure 5.20b shows the effect of included angle and length of the triangular patterns on the water collection rate under different conditions (Song and Bhushan 2019d). It shows that the water collection rate changes little with a change of included angle under conditions of fog, condensation and fog + condensation. However, the triangular length affected the water collection rate under all tested conditions. A shorter length benefits the water collection rate. Since a shorter distance requires less time to transport the droplets, although droplets being removed are small, the removal rate increases the collection rate. The water collection rate with combined fog and condensation is larger than under fog or condensation, separately. The water collection rate is around two times higher than the rate in the fog alone.

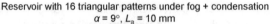

Reservoir with 16 triangular patterns under fog + condensation
$\alpha = 9°$, $L_a = 10$ mm

(a)

Effect of included angle on water collection rate
$L_a = 10$ mm

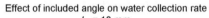
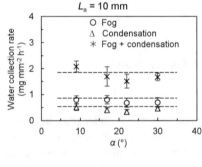

Effect of length on water collection rate
$\alpha = 9°$

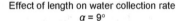
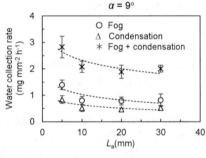

(b)

Fig. 5.20 **a** Representative optical images of the collected water in the reservoir by an array of triangular patterns under fog + condensation. **b** Water collection rate as a function of included angle and length of the triangular patterns (adapted from Song and Bhushan 2019d). The data of condensation at relative humidity of 85% from Song and Bhushan (2019b) and the data of fog from Song and Bhushan (2019c) was used

5.4.4 Summary

Water collection rates and transport ability of the triangular patterns under fog, condensation and combination of both were investigated. Figure 5.21 schematically shows the droplet collection and transport process under fog and/or condensation (Song and Bhushan 2019d). The droplet grows and coalesces. When the droplet grows large enough, it touches the superhydrophobic borders. A Laplace pressure gradient forms along the droplet and drives it to a wider position. The spontaneous drainage process continues for the new droplets fed into the triangular region (Song and Bhushan 2019d). The drained water can be collected in a reservoir before some of it is evaporated.

In condensation as compared to fog, the droplets grew faster, and the estimated water collection was increased to about two times that of the rate in just fog

Schematic showing the water collection and transport by a triangular
hydrophilic pattern with superhydrophobic borders

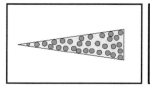 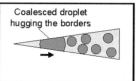 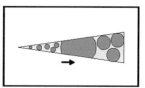

Small droplet formation   Coalesced droplet transport

Fig. 5.21 Schematic illustration of the water collection and transport process on a triangular hydrophilic pattern. Small droplets are deposited on the triangular hydrophilic area. When the smaller droplets grow and coalesce to bigger ones and touch the superhydrophobic borders. A Laplace pressure gradient exists along the droplet which is able to move the droplet. After the droplet is moved, the empty area will continue to collect small water droplets (adapted from Song and Bhushan 2019d)

condition. When the triangular pattern was subjected to both fog and condensation, the droplet collection and transport was faster than under fog or condensation alone. Mass measurement of the water collection using an array of triangular patterns showed that the condensation increased the water collection rate by approximately two times the rate in the fog alone. Included angle of the triangular geometry did not affect the water collection rate, whereas the shorter length increased the rate. The approach of combining fog and condensation can be used to develop efficient water collection systems (Song and Bhushan 2019d).

5.5 Results and Discussion—Water Condensation Studies Using Surfaces with Multistep Wettability Gradient

Wettability gradient on a surface can increase the droplet mobility. Figure 5.22 shows the droplet movement on a surface with heterogeneous wettability. Wettability gradient provides an unbalanced Young's force on both sides of the droplet as a driving force for directional transport of the droplet. If the droplet is placed on an inclined plane, the driving force is equal to the unbalanced Young's force experienced by a cross section of a droplet minus the gravitational force component along the surface given as (Adamson and Gast 1997; Israelachvili 2011; Bhushan 2018),

$$F = \gamma_{LA}(\cos\theta_B - \cos\theta_A)w - mg\sin\alpha \qquad (5.5)$$

where θ_A and θ_B are the contact angles of the adjacent steps on a surface with heterogeneous wettability, and w is the width, and m is the mass of the droplet.

Droplet movement on an inclined surface with heterogeneous wettability

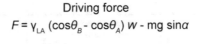

Driving force

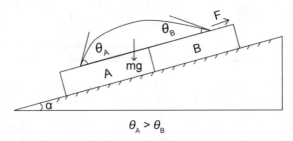

$\theta_A > \theta_B$

Fig. 5.22 Schematic of the droplet movement on an inclined surface with heterogeneous wettability

If the right step is more hydrophilic than the left step, the droplet will be driven toward the right side of the surface. Droplet on an inclined plane will move as long as the Young's force is larger than gravitational force component along the surface.

5.5.1   Single Droplet Experiments on Flat Surfaces

To understand the droplet transport mechanism on flat surfaces, single droplet experiments were carried out on hydrophilic surfaces with homogeneous and heterogeneous wettability. Droplets were deposited using a microsyringe at the left side of sample areas with a volume increment of 5 μL. Experiments were carried out on two hydrophilic surfaces with CA of 81° and 20°, and another with multistep wettability gradient (CA = 81°–20°). Selected optical images are shown in Fig. 5.23 (Feng and Bhushan 2020). A schematic of the rectangular array below the photo in the bottom row shows the wettability steps.

    The hydrophilic surface with CA = 81° did not have a wettability gradient, so there was no droplet transport. But due to the increased volume of the liquid and the limitation of the left edge, the droplet could expand to the right. When the volume of the droplet reached above 60 μL, the width of the surface could not hold the droplet. For another hydrophilic surface with CA = 20°, the droplet spread more on the surface due to the smaller contact angle, and the entire surface was covered with a thin water film when the droplet was at 10 μL. As the droplet volume increased, the water film became thicker. For the surface with the wettability gradient, the droplets rapidly spread over the surface as the droplets were deposited. As the contact angle of the next steps became smaller, the spread length of droplets

Effect of droplet volume on its location with an increment of 5 µL each time

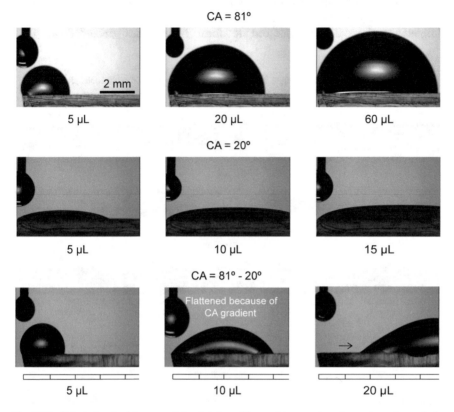

Fig. 5.23 Selected optical images of droplet deposition and transport on the rectangular surfaces with two wettabilities (CA = 20° or 81°) and with multistep wettability gradient (CA = 81°–20°) at different droplet volumes (adapted from Feng and Bhushan 2020)

gradually increased. As the droplet volume increased to greater than 20 µL, the droplets traversed more steps and after coalescing, the droplets were transported to the storage step (CA = 20°).

5.5.2 Rectangular Sample Patterns

Droplet condensation and transport were studied on samples with rectangular patterns with homogeneous and heterogeneous wettability. Samples with multistep wettability gradient (CA = 81°–20°) and two hydrophilic samples (CA = 81° and CA = 20°) were studied.

The selected optical images are shown in Fig. 5.24a (Feng and Bhushan 2020). In the hydrophilic surface with CA = 81°, at the beginning stage, the droplets coalesced and grew larger on the surface. As the condensation continued, more droplets condensed and coalesced to form larger ones (recorded as 1–7).

Fig. 5.24 **a** Selected optical images of condensed droplets on rectangular surfaces on a horizontal plane with two wettabilitities (CA = 20° or 81°) and with multistep wettability gradient (CA = 81°–20°). Arrows shown above some droplets represent movement of center of the droplet observed in videos. **b** Schematic illustration of droplet growth, coalescence and transport on rectangular surfaces with two wettability and with wettability gradient (adapted from Feng and Bhushan 2020)

Eventually, when the droplets were large enough to reach adjacent droplets, the coalescence of droplets formed larger ones (1 + 2 + 3, 4 + 5 and 6 + 7) at a coalescence time of 98 min. After the formation of the larger droplets, the center of the entire droplets did not change, which meant that the droplets were not transported after coalescence.

On another hydrophilic surface with CA = 20°, since the surface was more hydrophilic, the droplet spread more quickly. Droplets coalesced to form larger droplets more quickly. This took place in 13 min. Then in a short period of time, a water film was formed, and the entire process took only 32 min. No directional transport of any droplets was observed due to the high adhesion of the surface.

The initial process of condensation of droplets on the surface with multistep wettability gradient (CA = 81°–20°) was the same as on the two hydrophilic surfaces (CA = 81° and 20°). However, at 42 min after the droplets condensed, they were not uniform in size (recorded as 1–3). The right steps were more hydrophilic than the left steps, thus causing the droplets in the right steps to spread more. Between 51 and 63 min, as the droplets condensed and became larger, the wettability gradient provided an unbalanced force at the two sides of a droplet which acted as a driving force for directional movement of the droplet to the more hydrophilic step. Droplets 2 and 3 coalesced into large droplets (2 + 3) at 51 min. After that, the size of the large droplets was insufficient to coalesce with the droplets next to them due to the randomness of the droplets. Therefore, with the help of continuous condensation to create new droplets, they finally coalesced with large droplets (2 + 3) at 63 min and transported to the storage step, at which point the droplet transport was completed over the entire surface. The exposed area on the left steps could continue to have condensation and have new transport.

To better understand the transport mechanism of droplets on hydrophilic surfaces and the surfaces with wettability gradient, the growth, coalescence and transport of droplets are shown schematically in Fig. 5.24b (Feng and Bhushan 2020). For three different surfaces, first, droplets were condensed, which coalesced with other droplets to form larger droplets. For hydrophilic surfaces (CA = 81° and CA = 20°), the forces of the droplets were equal when coalescing. In the case of a more hydrophilic surface, and the droplets spread over larger area under the same volume. Therefore, coalescence was more likely to occur. In addition, the more hydrophilic surface takes the shorter time to form the water film. However, since the surface was more hydrophilic, the adhesion was higher. The droplets formed a very thin film of water that did not fall off when the surface was tilted. The surface with a wettability gradient provided directional transport of droplets. The CA on the left and right sides were different, which led to droplet transport.

Next, whether the droplets can climb on an inclined plane with multistep wettability gradient and overcome the gravitation forces was studied. The sample with multistep wettability gradient was placed on an inclined plane with an inclination angle of 5°. The droplets condensed, coalesced and transported upward. The selected optical images of droplets on samples sitting on flat and inclined planes are

Water condensation and transport on rectangular surfaces with wettability gradient (CA = 81° - 20°)
placed on two planes

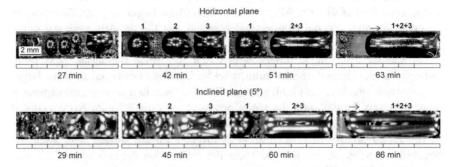

Fig. 5.25 Selected optical images of condensed droplets on rectangular surfaces with multistep wettability gradient (CA = 81°–20°), placed on a horizontal plane and another inclined at 5°, at different times. Arrows shown above some droplets represent movement of center of the droplet observed in videos (adapted from Feng and Bhushan 2020)

shown in Fig. 5.25 (Feng and Bhushan 2020). As expected, it took longer for droplets on an inclined plane to reach the storage step. However, the driving force due to heterogeneous wettability was high enough to overcome gravitational forces.

5.5.3 Triangular Patterns

To investigate the combined effects of Laplace pressure gradient and multistep wettability gradient, three samples with triangular patterns with different wettability were studied. They included two hydrophilic triangular surfaces (CA = 81° and CA = 20°) and one surface with wettability gradient (CA = 81°–20°). Selected optical images are shown in Fig. 5.26 (Feng and Bhushan 2020). Shapes of the moving droplets and their transport times were affected by surface wettability.

In the hydrophilic surfaces with CA = 81°, the condensed water droplets were relatively small at the onset of condensation. As condensation continues, the growing droplets began to coalesce into larger droplets. Ultimately, they were large enough to reach the boundary, triggering transport driven by Laplace pressure gradient. For example, after 36 min of condensation on the triangular sample, 5 large droplets (numbers 1–5) were formed to touch the boundary. At 48 min, droplets 1, 2 and 3 coalesced into one large droplet (1 + 2 + 3). In the next period of time of 3 min, large droplets (1 + 2 + 3) coalesced droplets 4 and 5, the droplet volume increased, and the Laplace pressure gradient could drive the droplets to the storage area. On another more hydrophilic surface with CA = 20°, the initial stage

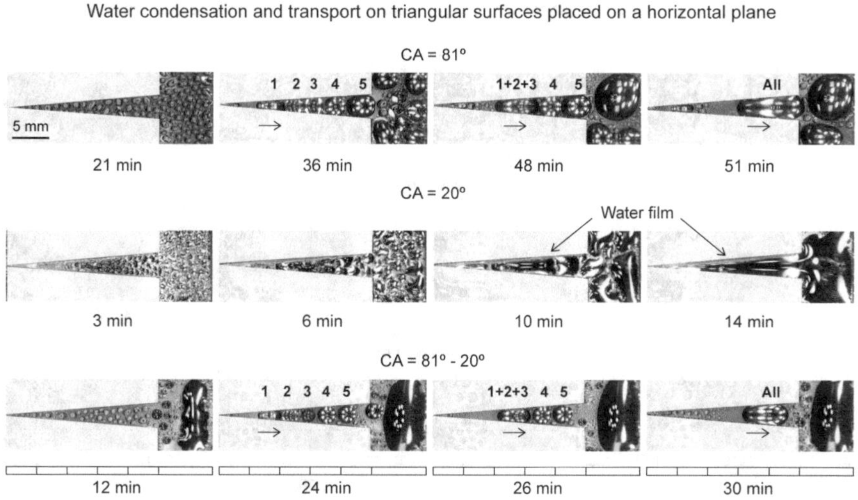

Water condensation and transport on triangular surfaces placed on a horizontal plane

Fig. 5.26 Selected optical images of condensed droplets on triangular surfaces placed on horizontal planes with two wettabilities (CA = 20° or 81°) and with multistep wettability gradient (CA = 81°–20°), at different times. Arrows shown above some droplets represent movement of center of the droplet observed in videos (adapted from Feng and Bhushan 2020)

of condensation was the same, but the whole process was very fast. At 14 min, all of the droplets on the entire surface coalesced to form a water film. Similar to the rectangular hydrophilic surface (CA = 20°), a thin water film was formed. The Laplace pressure gradient along the coalesced droplets could not overcome the high adhesion resulting in the transport of droplets.

The process of droplet condensation, coalescence and transport on two hydrophilic surfaces was similar. However, the triangular sample combined with the multistep wettability gradient and the Laplace pressure gradient had an increase in the coalescence rate of the droplets. By comparing the rectangular surface with wettability gradient, the entire transport time was shortened by approximately 50%. It was found that the key to rapid transport of droplets was the rapid coalescence of the droplets and the sufficient directional driving force so that the overall condensation and transport time was greatly reduced.

Next, whether the droplets can climb on an inclined plane with multistep wettability gradient and overcome the gravitation forces was studied. The three samples with triangular geometries placed on a plane inclined at 15° were studied. Selected optical images of droplets are presented in Fig. 5.27 (Feng and Bhushan 2020). As expected, it took longer for droplets on all surfaces on inclined planes to reach the storage steps. However, the driving forces in all samples were high enough to overcome gravitational forces.

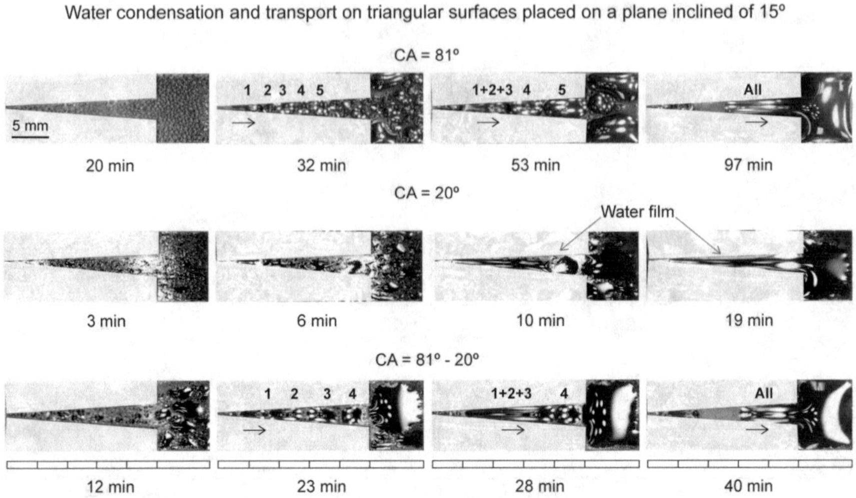

Fig. 5.27 Selected optical images of condensed droplets on triangular surfaces placed on a plane inclined at 15° with two wettabilities (CA = 20° or 81°) and with multistep wettability gradient (CA = 81°–20°), at different times. Arrows shown above some droplets represent movement of the center of the droplet observed in videos (adapted from Feng and Bhushan 2020)

5.5.4 Summary

Water condensation and transport on triangular surfaces with multistep wettability gradient were investigated. For comparison, experiments were also carried out on rectangular surfaces with various homogeneous wettability. On rectangular surfaces with wettability gradient, when the water in the ambient air was cooled to about 5°, it condensed, grew, and coalesced on the rectangle surface. When the coalesced droplets were driven by the wettability gradient, the droplets began to move and were transported to the storage step. The triangular surface combined with the wettability gradient and the Laplace pressure gradient had an increase in the coalescence rate of the droplets and the transport time was shorter than on the rectangular samples. Even on surfaces placed on inclined planes, the driving forces were high enough to transport droplets upward to the storage steps (Feng and Bhushan 2020).

5.6 Results and Discussion—Water Condensation Studies Using Nested Triangular

Patterns

To study the effect of nested triangular patterns on droplet transport, the water collection studies were conducted on hydrophilic triangular patterns with a single triangle and with two nested triangles.

5.6.1 Single Droplet Experiment

In order to study the relationship between the volume of the deposited droplet and the distance traveled, the single droplet experiments were carried out on patterns with a single triangle and with nested triangles. The selected optical images of patterns, after deposition of various droplet volumes, are shown in Fig. 5.28 (Bhushan and Feng 2020). For the pattern with a single triangle, after the droplet was deposited at the apex, the droplet touched the superhydrophobic dam, which

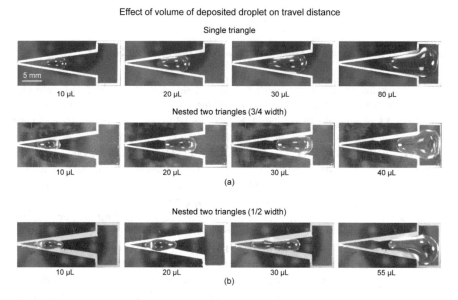

Fig. 5.28 Selected optical images after deposition of various droplet volumes (**a**) on patterns with a single triangle and with two nested triangles (with the initial width of the second triangle at the junction being equal to ¾ width of the first triangle), and (**b**) on a pattern with two nested triangles (with the initial width of second triangle at the junction being equal to ½ width of the first triangle) (adapted from Bhushan and Feng 2020)

triggered the droplet transport driven by the Laplace pressure gradient Fig. 5.28a (Bhushan and Feng 2020). The droplet stopped after some distance. Another droplet was placed at the current location of the existing droplet. However, the travel distance decreased with subsequent droplets. This was because the width of the pattern became larger, resulting in a lower Laplace pressure gradient (Bhushan 2018, 2019). Also, the liquid volume required to touch the borders became larger. For the pattern with two nested triangles, the droplet transport process at the length of 10 was similar to that of the pattern with a single triangle. However, the width of the nested pattern at the junction with second triangle at 10 mm was optimized to ¾ width of the first triangle. In the second triangle, the Laplace pressure gradient increased and the width was smaller, which accelerated the movement of the droplet. Therefore, in the case of nested triangles, a droplet with a volume of 40 μL reached the reservoir, instead of a droplet with a volume of 80 μL for a single triangle (Bhushan and Feng 2020).

When the width at the junction with the second triangle at 10 mm was only ½ width of the first triangle, the droplet was not able to move readily past the junction of the two triangles, as shown in Fig. 5.28b (Bhushan and Feng 2020). A droplet with a volume of 55 μL was needed to reach the reservoir, instead of a droplet with a volume of 40 μL for nested triangles with ¾ width of the first triangle at the junction. It was concluded that the droplets are not able to pass through the junction smoothly with ½ width of the first triangle, and continuous and rapid transport of the droplet could not be achieved (Bhushan and Feng 2020).

5.6.2 Water Condensation and Transport on Patterns

To compare the ability to transport condensed droplets, on a single triangle and nested triangles, water condensation and transport studies were carried out. The selected optical images are shown in Fig. 5.29a (Bhushan and Feng 2020). In the pattern with a single triangle, at the beginning stage, the droplets were relatively small and grew larger by coalescence. As the condensation continued, more droplets condensed and coalesced to form larger ones (numbered as 1 and 2). When the droplets were large enough to touch the superhydrophobic borders, the motion driven by the Laplace pressure gradient was triggered. The movement of droplet 1 occurred after 54 min. Then, coalescence of the droplets to form larger droplets (numbered 1, 2 and 3) occured at 74 min. Also, the tiny droplets continued to condense. Droplets 1 and 2 coalesced and moved forward, and coalesced with droplet 3, which reached the reservoir after 95 min (Bhushan and Feng 2020).

For the pattern with two nested triangles, at 54 min, because of a smaller enclosed area in the second triangle, the droplets could coalesce more quickly to form larger droplets (numbered 1, 2, 3 and 4). Since the width at 10 mm was smaller in the nested triangles, the driving force of the droplets increased which led

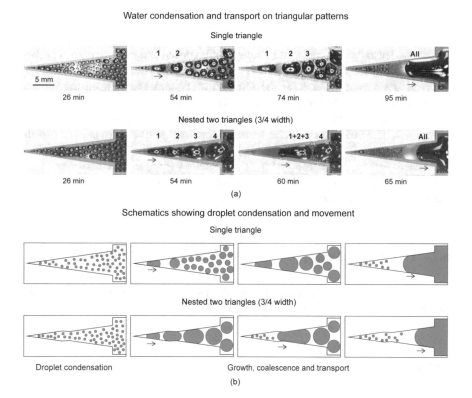

Fig. 5.29 a Selected optical images of condensed droplets on patterns with a single triangle and with two nested triangles at different times. Arrows shown above some droplets correspond to movement of the center of the droplet observed in videos, and **b** schematics of droplet nucleation, growth, coalescence and transport on patterns with a single triangle and with two nested triangles (adapted from Bhushan and Feng 2020)

to faster transport and coalescence into a larger droplet (1 + 2 + 3) after 60 min. It took 65 min to reach the reservoir in the nested triangles instead of 95 min in the case of a single triangle. Therefore, the nested pattern enables faster transport of droplets (Bhushan and Feng 2020).

To better understand coalescence and transport of droplets for the nested triangular patterns, schematics of growth, coalescence and transport of condensed droplets on the two patterns are shown in Fig. 5.29b (Bhushan and Feng 2020). First, the droplets are condensed, which coalesce with other droplets to form larger droplets. Once they touch the boundary, they start to move because of Laplace pressure gradient. The optimized pattern has two nested triangles, in which the Laplace pressure gradient was enhanced. At the same time, the width of the second triangle was smaller which required less liquid to touch boundaries and initiate transport. These led to faster transport of the droplets (Bhushan and Feng 2020).

5.6.3 Summary

Water condensation and transport studies were carried out on a pattern with two nested triangles for more efficient water collection. On the nested pattern, the initial width of the second triangle at the junction at 10 mm length from the tip was designed to be ¾ width of the first triangle in order to increase the Laplace pressure gradient. The data from the single droplet experiments and water condensation experiments showed that the water transport for the first triangular pattern was the same before and after the optimization of the pattern. But for the second triangle of the nested pattern, since the width of the second triangle was narrower at the junction of the two triangles, droplets traveled faster due to the higher Laplace pressure gradient (Bhushan and Feng 2020).

5.7   Design Guidelines for Water Harvesting Systems

Design of a bioinspired water harvesting 3D tower covered with triangular patterns will be presented in Chap. 6 (Bhushan 2020).

References

Adamson, A. V. and Gast, A. P. (1997), *Physical Chemistry of Surfaces*, sixth ed., Wiley, New York.

Alduchov, O. A. and Eskridge, R. E. (1996), "Improved Magnus Form Approximation of Saturation Vapor Pressure," *J. Appl. Meteorol.* **35**, 601–609.

Bhushan, B. (2018), *Biomimetics: Bioinspired Hierarchical-Structured Surfaces for Green Science and Technology,* third ed., Springer International, Cham, Switzerland.

Bhushan, B. (2019), "Bioinspired Water Collection Methods to Supplement Water Supply," *Philos. Trans. R. Soc. A* **377**, 20190119.

Bhushan, B. (2020), "Design of Water Harvesting Towers and Projections for Water Collection from Fog and Condensation," *Philos. Trans. R. Soc. A* **378**, 20190440.

Bhushan, B. and Feng, W. (2020), "Water Collection and Transport in Bioinspired Nested Triangular Patterns," *Philos. Trans. R. Soc. A* **378**, 20190441.

Bhushan, B. and Martin, S. (2018), "Substrate-Independent Superliquiphobic Coatings for Water, Oil, and Surfactant Repellency: An Overview," *J. Colloid Interface Sci.* **526**, 90–105.

Brochard, F. (1989), "Motions of Droplets on Solid Surfaces Induced by Chemical or Thermal Gradients," *Langmuir* **5**, 432–438.

Chaudhury, M. K. and Whitesides, G. M. (1992), "How to Make Water Run Uphill," *Science* **256**, 1539–1541.

Feng, W. and Bhushan, B. (2020), "Multistep Wettability Gradient in Bioinspired Triangular Patterns for Water Condensation and Transport," *J. Colloid Interface Sci.* **560**, 866–873.

Gurera, D. and Bhushan, B. (2019), "Designing Bioinspired Surfaces for Water Collection from Fog," *Philos. Trans. R. Soc. A* **377**, 20180269.

Israelachvili, J. N. (2011), *Intermolecular and Surface Forces*, third ed., Academic Press, Cambridge, Mass.

Lawrence, M. G. (2005), "The Relationship between Relative Humidity and the Dewpoint Temperature in Moist Air," *BAMS* **100**, 225-233.

Moran, M. J., Shapiro, H. N., Boettner, D. D., and Bailey, M. B. (2018), *Fundamentals of Engineering Thermodynamics*, Ninth ed., Wiley, New York.

Pruppacher, H. R. and Klett, J. D. (2010), *Microphysics of Clouds and Precipitation*, second ed., Springer, New York.

Rumble J. R. (2019), *CRC Handbook of Chemistry and Physics*, 100th Ed., CRC Press, Boca Raton, FL.

Sigsbee, R. A. (1969), *Nucleation*, Marcel Dekker, New York.

Song, D. and Bhushan, B. (2019a), "Water Condensation and Transport on Bioinspired Triangular Patterns with Heterogeneous Wettability at a Low Temperature," *Philos. Trans. R. Soc. A* **377**, 20180335.

Song, D. and Bhushan, B. (2019b), "Optimization of Bioinspired Triangular Patterns for Water Condensation and Transport," *Philos. Trans. R. Soc. A* **377**, 20190127.

Song, D. and Bhushan, B. (2019c). "Bioinspired Triangular Patterns for Water Collection from Fog," *Philos. Trans. R. Soc. A* **377**, 20190128.

Song, D. and Bhushan, B. (2019d), "Enhancement of Water Collection and Transport in Bioinspired Triangular Patterns from Combined Fog and Condensation," *J. Colloid Interface Sci.* **557**, 528–538.

Chapter 6
Commercial Applications, Projections of Water Collection, and Design of Water Harvesting Towers

6.1 Commercial Applications

Bioinspired large water harvesting towers and portable water harvesting towers are of commercial interest. Large water harvesting towers can be used to supply water to a community in arid regions. Portable units can be used to supply a home or a camper (Fig. 6.1) (adapted from image provided by Getty images) (Bhushan, 2020). Unlike in desalination, these units can be operated inland. Manufacturing and maintenance costs and energy consumption will be an important consideration for economic viability in these applications (Bhushan, 2020).

In addition, these towers can be used in various emergency and defense applications. Emergency applications, such as natural disasters, could benefit for short periods from portable units which could be dropped from air (Fig. 6.2) (Bhushan 2020). In emergency applications, human life is at stake, and cost is not an issue. Defense applications include military bases in combat zones. The cost of clean water in Forward Operating Bases (FOBs) in a combat zone such as in Afghanistan can be as much as US $350/gallon because of the dangers posed in transportation. Large towers can be installed at military bases located in deserts which can supply safe drinking water at a minimal cost, possibly couple of orders of magnitude lower than that of transported water (Fig. 6.3) (adapted from image provided by Jeffrey McGovern/U. S. Air Force) (Bhushan, 2020).

6.2 Projection of Water Collection Rates in Water Harvesting

Water collection rates for bioinspired water harvesters are projected and compared with the collection rates on flat surfaces and living species in deserts (Bhushan 2020). In arid deserts, based on some data presented in Chap. 2, water collection rates on flat surfaces are on the order of $2 \text{ L m}^{-2} \text{ day}^{-1}$ ($2 \text{ mg mm}^{-2} \text{ day}^{-1}$).

© Springer Nature Switzerland AG 2020
B. Bhushan, *Bioinspired Water Harvesting, Purification, and Oil-Water Separation*, Springer Series in Materials Science 299, https://doi.org/10.1007/978-3-030-42132-8_6

Personal use

Fig. 6.1 Schematic of a water harvesting tower placed in a home for personal use (adapted from Bhushan 2020)

The source of water is fog and condensation of water vapor mostly in nights. The bioinspired surfaces covered with conical arrays, based on data presented in Chap. 4 (Table 4.2), can provide higher collection rates from fog, about an order of magnitude larger than that of a flat hydrophilic surfaces. This means that bioinspired surfaces can collect water on the order of 20 L m^{-2} day^{-1}. These rates will be supplemented by condensation of water vapor. For an example of water collection of 20 L m^{-2} day^{-1}, a medium size tower covered with bioinspired surface with a surface area of 200 m^{2}, water collected would be about 4000 L day^{-1}. If the water consumption per capita is 50 L day^{-1} based on Chap. 1 (Table 1.1), a tower can provide sufficient water for about twenty families with 4 people in each family. The data are summarized in Table 6.1 (Bhushan 2020).

6.3   Design of Water Harvesting Towers

For scaleup, large 3D towers can be designed for high water collection. Inspiration was taken from beetles, grass, cactus, and spider silk. Slotted tower surfaces were selected to reduce swirl by intercepting wind. Figures 6.4 and 6.5 show schematics

Disaster struck area

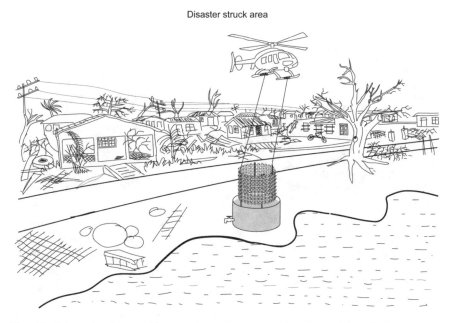

Fig. 6.2 Schematic of a portable water harvesting tower being dropped by a helicopter in an emergency zone on an island (adapted from Bhushan 2020)

of water harvesting towers covered with triangular patterns or conical array, respectively (Bhushan 2020). In the case of cones, heterogeneous wettability and grooves are used and are mounted at 45° to benefit from gravitational forces. Grooves on the ribs are provided to facilitate water flow down to the reservoir. Water droplets falling on the cones underneath provide cascading effect to accelerate fluid flow. In the case of triangular patterns, patterns are placed with tip pointed outward on the shims oriented horizontally to the vertical ribs. The collected water flows down the grooved ribs to the reservoir. In addition, water droplets will be deposited on the slot edges and will eventually fall down to the reservoir (Bhushan 2020).

For condensation of water vapor, the cooling process of the cones or shims with triangular patterns in the towers is provided by the low temperatures in desert nights. If additional cooling is required, one could take advantage of the renewable daytime solar energy. In the field, the water harvesting systems will get covered by dust. They will require filters to keep the active collection area clean. Filters may require high pressure air, which again can be provided by solar energy. The towers could be operated with near zero energy (Bhushan 2020).

Military base in desert

Fig. 6.3 Schematic of a water harvesting tower placed on a military base in the combat zone in the desert (adapted from Bhushan 2020)

Table 6.1 Projected water collection rates from fog. These rates will be supplemented by condensation of water vapor (adapted from Bhushan 2020)

Type of surface	Approximate water collection rate (L m^{-2} day)
Flat surface in arid desert (Chap. 2)	2
Bioinspired surfaces covered with cones or triangular patterns (Chaps. 4 and 5)	20[a] ($\times$10 increase)

[a]A medium size bioinspired tower with a surface area of 200 m^2 can collect 4000 L day^{-1}

6.4 Operational and Maintenance Cost

The operational and maintenance cost needs to be minimum for field use, particularly for consumer applications. As a reference, the cost of bottled water can range from US $0.20 to $1.50 in different countries. One liter of desalinated water costs about 1 cent US. The target for bioinspired designs for consumer applications is less than 2 cents US. The data is summarized in Table 6.2 (Bhushan 2020). Given that operating cost for water harvesting tower is negligible and only maintenance cost exists, the target cost is achievable.

It should be noted that cost sensitivity is dependent upon the application. For emergency and defense applications, the cost is not an issue. Bioinspired approaches become very attractive.

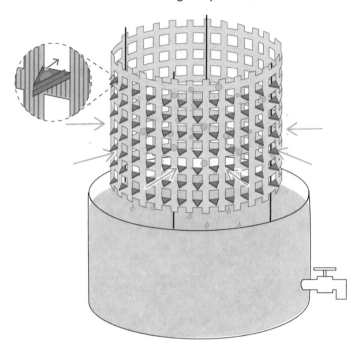

Bioinspired water harvesting 3D tower
with triangular patterns

Fig. 6.4 Schematic of a bioinspired water harvesting 3D tower covered with triangular patterns (adapted from Bhushan 2020)

6.5   Scaleup and Commercialization Issues

Significant investments are needed for scale up designs, technology transfer and product launch. Major innovation prizes have a long record of spurring innovation in private, public and philanthropic sectors (Bhushan 2015). They have potential to facilitate quantum jump in technologies by attracting investors across the spectrum including those out-of-discipline. Although total investments by various competitive teams may be several fold they also only pay for success. In 2016, X-Prize Foundation announced US $1.75 M Water Abundance XPrize, a two year competition to develop energy-efficient technologies that harvest water from thin air. The goal was to develop a water harvesting device that can deliver 2000 L of water per day at a cost of no more than 2 cents US using only renewable energy. In Oct 2018, the Skysource/Skywater Alliance was awarded a grand prize of US $1.5 M.

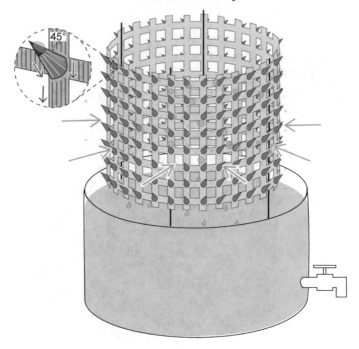

Fig. 6.5 Schematic of a bioinspired water harvesting 3D tower covered with conical array (adapted from Bhushan 2020)

Table 6.2 Water collection costs for various approaches (adapted from Bhushan 2020)

Water collection approach	Approximate cost (L^{-1})
Bottled water (for reference)	US $0.20 to $1.50
Desalination	1 cent US
Bioinspired design target	2 cents US (achievable)

References

Bhushan, B. (2015), "Perspective: Science and Technology Policy—What is at Stake and Why Should Scientists Participate?" *Sci. Publ. Policy* **42**, 887–900.

Bhushan, B. (2020), "Design of Water Harvesting Towers and Projections for Water Collection from Fog and Condensation," *Phil. Trans. R. Soc. A* **378**, 20190440.

Chapter 7
Bioinspired Water Desalination and Water Purification Approaches Using Membranes

As discussed in Chap. 1, 97.5% of water is saline water, therefore water desalination is increasingly important in some parts of the world. However, water desalinization remains an energy intensive process and prohibitively expensive. In addition, water contamination from human activity affects clean water supply. Water purification from all contaminants is important (Brown and Bhushan 2016; Bhushan 2018).

For water desalination and water purification, a commonly used commercial technique is reverse osmosis. Normally, if two aqueous solutions with varying solute concentrations are placed either side of a semipermeable membrane, water will move through the membrane from a region of low solute concentration to a region of higher solute concentration in an attempt to equalize the concentrations, Fig. 7.1 (Brown and Bhushan 2016). This process is known as osmosis and the tendency for a solution to take in water is defined as the osmotic pressure. In reverse osmosis (RO), external pressure is applied to overcome this osmotic pressure, preventing the flow of water into the region of higher solute concentration. Additional pressure above the osmotic pressure will instead cause water to move into the region of lower solute concentration. The requirement for this applied pressure means that separation via a RO membrane can be energy intensive, consuming at least 2 kWh m^{-3} (Lee et al. 2011), whereas the theoretical minimum energy required for desalination should be around 1 kWh m^{-3} (Elimelech and Phillip 2011). This excess pressure is due to the low permeability of the membranes involved. A membrane with low water permeability will require additional applied pressure, above that required to balance the osmotic pressure in order to result in reasonable water fluxes (Elimelech and Phillip 2011).

For water purification from various contaminants, various techniques are used, including adsorbents such as activated carbon (Pollard et al. 1992), biomaterials (Crini 2005), and zeolites (porous minerals) (Wang and Peng 2010) to remove organics from wastewater (Brown and Bhushan 2016). However, adsorbents are

© Springer Nature Switzerland AG 2020
B. Bhushan, *Bioinspired Water Harvesting, Purification, and Oil-Water Separation*, Springer Series in Materials Science 299, https://doi.org/10.1007/978-3-030-42132-8_7

Osmosis and reverse osmosis

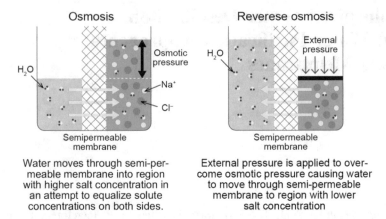

Fig. 7.1 Two water purification approaches—osmosis and reverse osmosis. In osmosis, water travels across a semipermeable membrane to equalize solute concentrations. In reverse osmosis, external pressure is applied to prevent water from traveling into regions of higher solute concentration. Additional pressure, instead, causes water to travel into regions of lower solute concentrations (adapted from Brown and Bhushan 2016)

inefficient due to contaminant removal and regeneration of the adsorbent. Ultraviolet (UV) treatments have been used for water purification (Shannon et al. 2008; Li et al. 2019).

Micro- and nanoporous membranes are commonly used for separation of salt as well as contaminants (Brown and Bhushan 2016). There are two mechanisms that dictate water transport through a membrane, both of which can occur together. One mechanism is solution-diffusion, where water molecules dissolve into the membrane and diffuse through the membrane to desorb from the other side. Another mechanism is pore flow, where the size of the pores is smaller than the contaminant being removed (Sengur-Tasdemir et al. 2016). Microporous membranes are typically made of polymers and ceramic materials. Nanoporous membranes are typically made of graphene (Surwade et al. 2015) or etched silicon (Cavallo and Lagally 2010).

Porous materials are classified into three main categories depending upon their pore size. A summary of the various pore sizes, including solutes that can be targeted in this size range, examples in nature, and their bioinspired equivalents is provided in Fig. 7.2 and Table 7.1 (Brown and Bhushan 2016). Microporous materials contain pore diameters <2 nm, which are required for desalination. Ocean water (and other water sources) also contains many contaminants other than salt which are typically many orders of magnitude larger than the inorganic ions found in salt. When salt removal is not required, membranes with larger pores would likely be sufficient for these applications. In addition to separation applications, ordered macroporous (pore diameters 2–50 nm) or mesoporous (pore diameters >50 nm) materials could also find use in ion exchange (Davis 2002), catalysis (Taguchi and Schüth 2005) and battery technology (Esmanski and Ozin 2009).

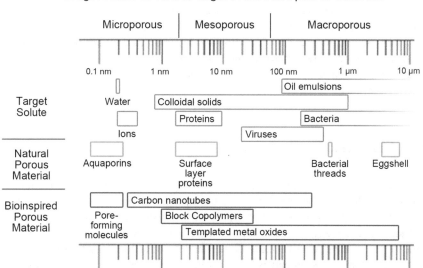

Fig. 7.2 Comparison of length scales of target solutes, natural porous materials, and bioinspired porous materials (adapted from Brown and Bhushan 2016)

Table 7.1 Examples of target solute, natural materials, and bioinspired equivalents sorted by membrane pore size (adapted from Brown and Bhushan 2016)

Pore size	Target solute	Natural porous material	Bioinspired porous material
Macropores (>50 nm)	Bacteria (0.2–750 μm) Oil emulsions (0.1–10 μm)	Eggshell, bacterial threads	Carbon nanotubes, templated silica
Mesopores (2–50 nm)	Colloidal solids (0.01–1 μm) Viruses (20–400 nm) Proteins (2–10 nm)	Surface layer proteins	Carbon nanotubes, self-assembled block copolymers
Micropores (0.2–2 nm)	Inorganic ions (0.2–0.4 nm) Water (0.2 nm)	Aquaporins	Carbon nanotubes, amphiphilic dipeptides, cyclic peptides, crown ethers

Living nature provides examples of membranes with higher separation rates and permeabilities than man-made equivalents. A steel wire mesh or nanofibrous polymer membrane can be coated with bioinspired material, which allows water flow but repels organic contaminants, which can be used for water purification and oil-water separation. These coated meshes can be used for water purification (Brown and Bhushan 2015, 2016; Bhushan 2018, 2019).

In this chapter, an overview of bioinspired membranes for water desalination and water purification is presented. Coated steel wire mesh and cotton for oil-water separation techniques and UV light-stimuli materials for water purification will be described in Chap. 9.

7.1 Multi-cellular Structures

Bioinspired microporous membranes have been produced by templating techniques (Brown and Bhushan 2016). One early example is the use of bacterial threads to form ordered silica macrostructures. This was achieved by Davis et al. (1997) by first slowly extracting a thread from a culture of *B. subtilis*, which caused the multicellular filaments to become aligned into a collection of hexagonally closed packed cylinders. The thread was then dipped into an aqueous sol comprising colloidal silica nanoparticles. As the thread was dried, the nanoparticles coated the thread, after which the organic material was then removed via heating to result in pore widths of 0.5 μm.

Various other examples of templating membranes are shown in Fig. 7.3 (Brown and Bhushan 2016). Template mineralization has been performed using the internal shell of *Sepia officinalis* or cuttlefish, Fig. 7.3a (Ogasawara et al. 2000). This cuttlefish bone consists of a highly organized structure comprised of calcium carbonate on a chitin framework. By removing the calcium carbonate from the framework, the chitin can be remineralized using an aqueous sodium silicate solution. The chitin template can then be removed by calcining. A similar technique utilized potato starch gels as the template to create silicalite (one of the forms of silicon dioxide) nanoparticle thin films with hierarchical porosity, Fig. 7.3b (Zhang et al. 2002). It was found that the pore size could be varied from 0.5 to 50 μm by altering the amount of starch in the suspension.

A method for creating template metal oxide membranes involved the use of eggshells, Fig. 7.3c (Yang et al. 2002). Eggshells contain membranes with a macroporous network of interwoven fibers with typical pore sizes of 5 μm. The membrane surface consists of amines and carboxylic acid groups capable of interacting with titania precursors. The eggshell was first dipped in a tetra-n-butyl titanium solution before being transferred to a propanol/water mixture, resulting in the hydrolysis and condensation of titania onto the surfaces of the eggshell membrane. After removal of the organic material through heating, a titania porous network remained. For smaller pore sizes, wood cellular structures have been combined with surfactant to create hierarchically ordered porous material, Fig. 7.3d (Shin et al. 2001, 2003). The wood was soaked in a solution containing surfactant and silicate. The silicate penetrated the cell walls and was hydrolyzed and condensed onto the internal structures of the wood. The surfactant was added and

Various techniques to fabricate multicellular-inspired membranes

(a) Cuttlebone templated chitin–silica
(Ogasawara et al., 2000)

(b) Starch templated silicalite
(Zhang et al., 2002)

(c) Eggshell membrane templated titania
(Yang et al., 2002)

(d) Wood cellular templated silica
(Shin et al., 2001)

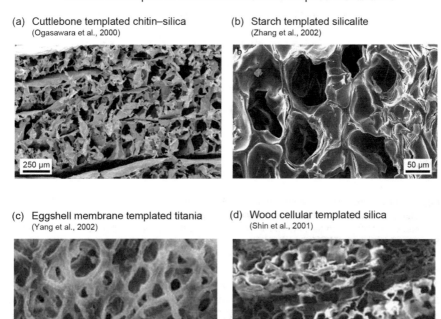

Fig. 7.3 Summary of various templating techniques for fabricating multicellular-inspired membranes: **a** cuttlebone-templated silica featuring a highly organized internal structure (adapted from Ogasawara et al. 2000), **b** starch-templated silicalite where pore size is varied by altering the amount of starch (adapted from Zhang et al. 2002), **c** eggshell-templated titania featuring interwoven fibres and typical pore sizes of 5 μm (adapted from Yang et al. 2002), and **d** wood cellular-templated silica where changing the species of wood can alter the pore structure (adapted from Shin et al. 2001; Brown and Bhushan 2016)

became incorporated into the silica network, forming nanoporous channels necessary for the evacuation of the organic matter during calcination. It was found that using different tree samples as the template resulted in different pore diameter, volume, and surface area. Pore sizes as well as 2.5–3 nm were achieved.

7.2 Aquaporins

Aquaporins are pore-forming proteins found in living cells (Agre et al. 1993; Brown and Bhushan 2016). These proteins have been found to facilitate a rapid and selective transport of water while blocking the passage of ionic species. They do so

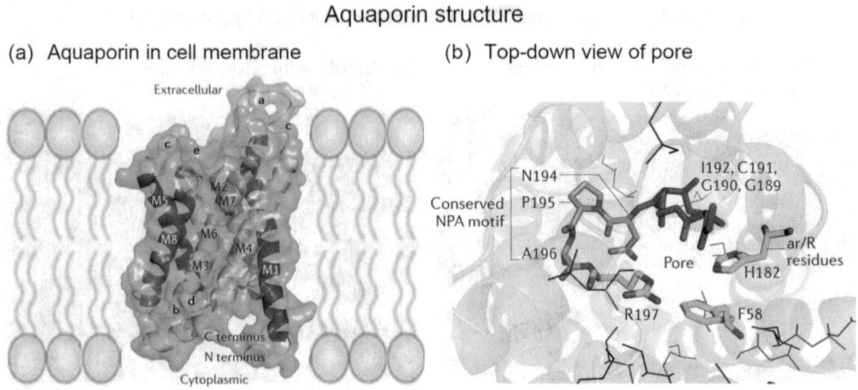

Aquaporin structure

(a) Aquaporin in cell membrane   (b) Top-down view of pore

Fig. 7.4 Schematic showing **a** aquaporin in cell membrane and **b** top-down view of pore and surrounding amino acid residues that contain hydrogen bond acceptors and donors to facilitate transport of water through the pore and act as selectivity filters (adapted from Verkman et al. 2014; Brown and Bhushan 2016)

by folding in such a way that they form hourglass shapes, with a pore running down the center of each. At its narrowest point, the width of the pore is defined by a certain peptide sequence, which allows different aquaporins to exhibit different pore sizes. Several clusters of amino acids containing polar side chains create a hydrophilic surface and, as hydrogen bond acceptors and donors, facilitate transport of water through the pore, Fig. 7.4 (Verkman et al. 2014).

Bioinspired aquaporin-based membranes exhibit superior water flux and solute rejection when compared to conventional membranes (Zhao et al. 2012). The membranes have been fabricated by combining the aquaporin protein (embedded within amphiphilic molecules, such as lipids) with a polymer support structure (Habel et al. 2015). Other approaches to fabricate nanoporous structures are being pursued. Various examples are described next (Brown and Bhushan 2016).

7.2.1 Pore-Forming Molecules

Figure 7.5 shows various techniques to fabricate aquaporin-inspired membranes by using other molecules that can mimic the pore-forming nature of the proteins (Brown and Bhushan 2016). Figure 7.5a shows the use of cyclic peptides (Ghadiri et al. 1993). The ring-shaped peptide consists of eight amino acids with alternating chirality (D or L) such that all the amide functionalities lie perpendicular to the plane of the ring. This allows for intermolecular hydrogen bonding between rings resulting in a tubular structure that was found to reach up to hundreds of nanometers in length. The alternating chirality also ensures that the amino acid side groups (R groups) all lie on the outside of the ring structure to result in tubes with an internal diameter of 0.7–0.8 nm. The authors suggested such structures could find use in catalysis, electronics, or separations; more recent studies have involved creating designs more compatible with polymeric membrane processing (Hourani et al. 2011).

Figure 7.5b shows the use of dendritic (branched) dipeptides that can self-assemble, either in solution or when cast in a film, into helical pores (Percec et al. 2004). Unlike the cyclic peptides mentioned above, the pores formed from the dendritic dipeptides contain amino acid side groups on their interior. This was found to result in a hydrophobic channel, which is favorable for fast transport of water with high selectivity. It is also possible to alter the pore structure by replacing sequences in the dipeptides. For instance, pore sizes can be varied from 0.2 to 2.4 nm by altering the peptide apex or branches. Proton transport measurements concluded that these pores are functional.

Figure 7.5c shows the use of crown ethers functionalized with ureido groups (Cazacu et al. 2006). This molecule can self-organize into tubes via hydrogen

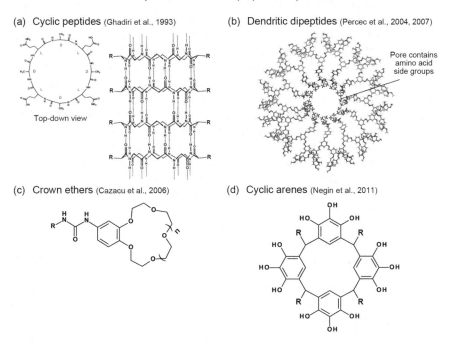

Fig. 7.5 Summary of various techniques for fabricating aquaporin-inspired membranes: **a** cyclic peptides containing a pore with a 0.7 nm internal diameter (adapted from Ghadiri et al. 1993), **b** dendritic peptides featuring hydrophobic side groups within the pore for fast transport of water (adapted from Percec et al. 2007), **c** crown ethers where the size of the ring n can be altered (adapted from Cazacu et al. 2006), and **d** cyclic arenes, which form hexamers that go on to form a channel with pores 1.8 nm in diameter (adapted from Negin et al. 2011; Brown and Bhushan 2016)

bonding of these groups. By changing the number of ether groups within the ring (n), the pore size can be altered. These molecular channels were found to self-assemble within an existing lipid membrane, though conductivity studies of species through the channels gave inconsistent results.

Figure 7.5d shows the use of calixarenes which are cyclic macromolecules featuring aromatic groups (Negin et al. 2011). These molecules were found to organize into nanotubes due to the interlocking of the side chains (R groups), to form a cyclic hexamer featuring six molecules in a doughnut shape. It is these hexamers that are then found to form the channel, with pore sizes of 1.8 nm.

Pore-forming molecules are typically incorporated into lipid membranes, however, they may not be robust (Brown and Bhushan 2016).

7.2.2 Carbon Nanotubes

Carbon nanotubes (CNT) have been used to replicate the role of aquaporins non-biological channels (Brown and Bhushan 2016). Although they are intrinsically hydrophobic, water transport has been found to be possible through such confined channels (Hummer et al. 2001). The water transport occurs to the atomically flat walls and the formation of hydrogen bonds between adjacent water molecules inside the tube both leading to a smooth energetic landscape for water travelling inside the nanotubes (Corry 2008). Examples of CNTs have been successfully utilized as filters for the removal of contaminants of various sizes.

Srivastava et al. (2004) grew aligned multi-walled carbon nanotubes (MWCNT) on a hollow carbon cylinder through a spray pyrolysis technique with the inner diameter of the nanotubes of 10–12 nm. The material was found to remove *E. coli* (2–5 μm), *Staphylococcus aureus* (1 μm), and poliovirus (ca. 25 nm) from water. However, smaller diameters are necessary for blocking inorganic ions. Holt et al. (2006) grew narrower double-walled carbon nanotubes (DWCNT) on a pitted silicon chip via catalyzed chemical vapor deposition and were encapsulated with vapor-deposited silicon nitride, Fig. 7.6 (Brown and Bhushan 2016). Reactive ion etching was used to open the ends of the nanotubes, and size exclusion measurements and transmission electron microscope (TEM) images determined the average inner diameter of the nanotubes to be 1.6 nm. Despite the small pore size, the DWCNT membranes were found to have superior permeability compared to conventional polymer membranes. When negatively charged functionality is added to the entrance of the DWCNTs, the membranes exhibit good ion rejection, a product of interactions between particle charge and surface charge (Fornasiero et al. 2008). Fabrication of aligned DWCNTs is expected to be expensive.

Fig. 7.6 **a** Schematic showing synthesis of aligned carbon nanotube membranes, **b** SEM image of carbon nanotubes atop silicon substrate, and **c** distribution of pore sizes as determined by TEM measurements (adapted from Holt et al. 2006; Brown and Bhushan 2016)

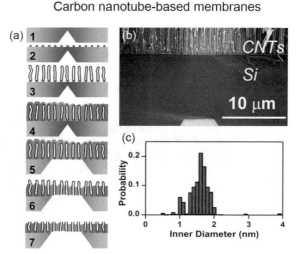

7.2.3 Self-assembled Block Copolymers

In addition to carbon nanotubes, self-assembled block copolymers have been used to replicate the role of aquaporins (Brown and Bhushan 2016). Self-assembly of block copolymers occurs due to inherent differences between the constituent blocks, leading to phase separation. Selection of appropriate conditions (concentration, solvent, drying times, etc.) can dictate this separation to result in various structures, including hexagonally packed cylindrical phases (Liu and Ding 1998; Phillip et al. 2010). Selective dissolution or etching of one of the blocks then results in a microporous structure comprising the remaining polymeric material (Liu and Ding 1998). This technique is able to produce pore sizes of 8–30 nm.

Figure 7.7 shows an example with polystyrene-*b*-poly(4-vinylpyridine) (PS-*b*-P4VP) (Peinemann et al. 2007). The diblock copolymer solution typically comprises a dimethylformamide (DMF) and tetrahydrofuran (THF) solvent mixture. DMF is more selective towards P4VP, while THF is more selective towards PS. The more volatile nature of THF leads to a concentration-induced phase separation of the block copolymer resulting in a less swollen PS matrix surrounding highly swollen, cylindrical P4VP domains. Immersing the drying polymer film into a non-solvent (in this case water) leads to solvent exchange in the swollen P4VP domains due to water possessing a higher compatibility for P4VP than DMF. The water is guided through the interconnected P4VP domains and leads to the solidification of the water-insoluble polystyrene matrix. The eventual shrinkage during drying of the swollen P4VP domains results in the formation of pores. Size exclusion experiments determined the pores have an effective diameter of 8 nm, with an 82% rejection of albumin (Brown and Bhushan 2016).

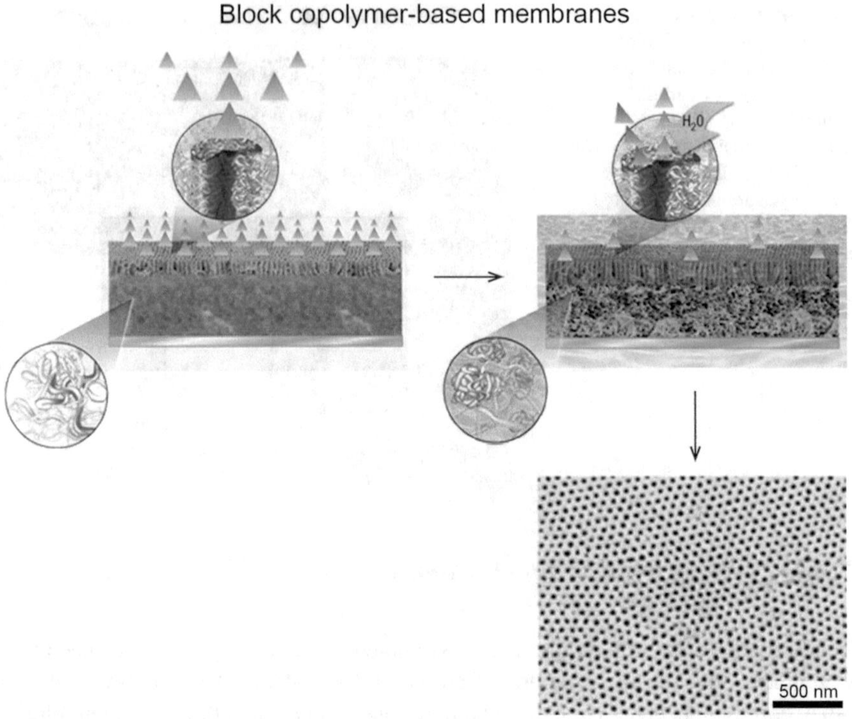

Block copolymer-based membranes

Fig. 7.7 Schematic of a block copolymer-based membrane showing nonsolvent-induced phase separation of polystyrene-b-poly(4-vinylpyridine) from THF/DMF solvent mixture. Concentration-induced phase separation results in a less swollen PS matrix surrounding highly swollen P4VP domains. Immersion in water results in solvent exchange within P4VP domains and solidification of water-insoluble PS matrix. Eventual shrinking during drying of P4VP leads to formation of pores. SEM of porous block copolymer membrane (adapted from Peinemann et al. 2007; Brown and Bhushan 2016)

7.3 Dual pH- and Ammonia-Vapor-Responsive Electrospun Nanofibrous Polymer Membranes with Superliquiphilic/Phobic Properties

Ma et al. (2017) fabricated a dual pH- and ammonia-vapor-responsive membrane by solution dip-coating an electrospun polyimide (PI) nanofibrous membrane. PI-based nanofibrous membranes are considered to be one of the most versatile membrane types due to their excellent mechanical strength, flexibility and stability and have been widely used. The electrospun PI nanofibrous membranes were dip-coated in a decanoic acid (DA)-TiO_2 mixture and a silica nanoparticles (SNPs) pre-gel solution, followed by high-temperature annealing. The silica nanoparticles embedded in the fiber surface increase the membrane surface roughness and hence

Electrospun nanofibrous membrane

 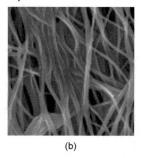 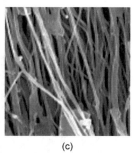

(a)   (b)   (c)

Fig. 7.8 SEM images for **a** the pristine PI nanofibrous membrane with an average diameter of approximately 300 nm, **b** PI nanofibrous membrane treated with DA-TiO$_2$ sol and **c** the PI nanofibrous membrane treated with decanoic acid (DA)-TiO$_2$ sol and silica nanoparticles (SNPs) to enhance surface roughness (adapted from Ma et al. 2017)

hydrophilicity/hydrophobicity. Figure 7.8 shows scanning electron microscope (SEM) images of the SNPs/DA-TiO$_2$/PI membrane at various stages of fabrication.

Ma et al. (2017) measured the pH-responsive wettability of the nanofibrous membranes. The membrane was superhydrophobic/superoleophilic underwater with acidic conditions (measured at a pH 6.5) but was superhydrophilic/superoleophobic in basic conditions (measured at a pH 12). They reported that the surface wettability transition could also be induced by exposure to ammonia vapor. Due to the tunability of the membrane surface wettability with pH- and ammonia vapor, this membrane could be used for oil-water separation under various conditions. Ma et al. (2017) performed oil-water mixture separation experiments at a range of pH and exposure to ammonia vapor. They demonstrated feasibility of the membranes.

These membranes can be used for oil-water separation and water purification to remove organic contaminants with high purity.

7.4 Summary

Desalination via reverse osmosis is the most commonly used water purification technique. Porous membranes being used have poor water permeabilities, requiring additional applied pressure above that required to overcome osmotic pressure and achieve reasonable water fluxes. Bioinspired membranes typically include a channel-forming component, such as carbon nanotubes and self-assembled block copolymers, Fig. 7.9 (Brown and Bhushan 2016). Many examples can be found to exhibit higher permeabilities compared to conventional membranes. For instance, even when added in very small percentages, the incorporation of CNTs or aquaporins into commercial RO membranes results in higher water flux and lower costs

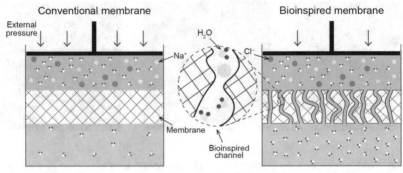

Fig. 7.9 Comparison of reverse osmosis membranes used in desalination. The semi-permeable nature of the current membranes results in the application of additional pressure to achieve reasonable water fluxes. Bioinspired membranes have much higher permeabilities and require less applied pressure (adapted from Brown and Bhushan 2016)

(Elimelech and Phillip 2011). However, there are some drawbacks to using CNTs for water purification. Creating aligned SW/DWCNT via chemical vapor deposition can be expensive. In addition, the CNTs must be further functionalized if they are to display the ion rejection efficiencies necessary for desalination and these modifications may alter or destroy the nanotube walls. Self-assembled block copolymers offer advantages over CNTs, such as ease of fabrication, there are also some drawbacks. Currently, the pore size achievable through this method is still quite large, with pores <2 nm yet to be fabricated (Brown and Bhushan 2016).

References

Agre, P., Sasaki, S., and Chrispeels, M. J. (1993), "Aquaporins: a Family of Water Channel Proteins," *Am. J. Physiol.* **265**, F461.

Bhushan, B. (2018), *Biomimetics: Bioinspired Hierarchical-Structured Surfaces for Green Science and Technology*, third ed., Springer International, Cham, Switzerland.

Bhushan, B. (2019), "Bioinspired Oil-water Seperation Approaches for Oil Spill Clean-up and Water Purification," *Phil. Trans. R. Soc. A* **377**, 20190120.

Brown, P. S., and Bhushan, B. (2015), "Bioinspired, Roughness-Induced, Water and Oil Super-philic and Super-phobic Coatings Prepared by Adaptable Layer-by-Layer Technique," *Sci. Rep.* **5**, 14030.

Brown, P. S. and Bhushan, B. (2016), "Bioinspired Materials for Water Supply and Management: Water Collection, Water Purification and Separation of Water from Oil," *Phil. Trans. R. Soc. A* **374**, 20160135.

Cazacu, A. Tong, C., van der Lee, A., Fyles, T. M., and Barboiu, M. (2006), "Columnar Self-Assembled Ureido Crown Ethers: An Example of Ion-Channel Organization in Lipid Bilayers," *J. Am. Chem. Soc.* **128**, 9541–9548.

Cavallo, F. and Lagally, M. G. (2010), "Semiconductors Turn Soft: Inorganic Nanomembranes," *Soft Matter* **6**, 439–455.

Crini, G. (2005), "Recent Developments in Polysaccharide-based Materials used as Adsorbents in Wastewater Treatment," *Prog. Polym. Sci.* **30**, 38–70.

Corry, B. (2008), "Designing Carbon Nanotube Membranes for Efficient Water Desalination," *J. Phys. Chem. B* **112**, 1427–1434.

Davis, M. E. (2002), "Ordered Porous Materials for Emerging Applications," *Nature* **417**, 813–820.

Davis, S. A., Burkett, S. L., Mendelson, N. H., and Mann, S. (1997), "Bacterial Templating of Ordered Macrostructures in Silica and Silica-surfactant Mesophases," *Nature* **385**, 420–423.

Elimelech, M. and Phillip, W. A. (2011), "The Future of Seawater Desalination: Energy, Technology, and the Environment," *Science* **333**, 712–717.

Esmanski, A. and Ozin, G. A. (2009), "Silicon Inverse-Opal-Based Macroporous Materials as Negative Electrodes for Lithium Ion Batteries," *Adv. Funct. Mater.* **19**, 1999–2010.

Fornasiero, F., Park, H. G., Holt, J. K., Stadermann, M., Grigoropoulos, C. P., Noy, A., and Bakajin, O. (2008), "Ion exclusion by sub-2-nm carbon nanotube pores," *Proc. Natl. Acad. Sci.* **105**, 17250–17255.

Ghadiri, M. R., Granja, J. R., Milligan, R. A., McRee, D. E., and Khazanovich, N. (1993), "Self-assembling Organic Nanotubes Based on a Cyclic Peptide Architecture," *Nature* **366**, 324–327.

Habel, J., Hansen, M., Kynde, S., Larsen, N., Midtgaard, S. R., Jensen, G. V., Bomholt, J., Ogbonna, A., Almdal, K., Schulz, A., and Hélix-Nielsen, C. (2015), "Aquaporin-based Biomimetic Polymeric Membranes: Approaches and Challenges," *Membranes* **5**, 307–351.

Holt, J. K., Park, H. G., Wang, Y., Stadermann, M., Artyukhin, A. B., Grigoropoulos, C. P., Noy, A., and Bakajin, O. (2006), "Fast Mass Transport through Sub-2-Nanometer Carbon Nanotubes," *Science* **312**, 1034–1037.

Hourani, R., Zhang, C., van der Weegen, R., Ruiz, L., Li, C., Keten, S., Helms, B. A., and Xu, T. (2011), "Processable Cyclic Peptide Nanotubes with Tunable Interiors," *J. Am. Chem. Soc.* **133**, 15296–15299.

Hummer, G., Rasaiah, J. C., and Noworyta, J. P. (2001), "Water Conduction Through the Hydrophobic Channel of a Carbon Nanotube," *Nature* **414**, 188–190.

Lee, K. P., Arnot, T. C., and Mattia, D. (2011), "A Review of Reverse Osmosis Membrane Materials for Desalination—Development to Date and Future Potential," *J. Membr. Sci.* **370**, 1–22.

Li, F., Kong, W., Bhushan, B., Zhao, X., and Pan, Y. (2019), "Ultraviolet–driven Switchable Superliquiphobic/superliquiphilic Coating for Separation of Oil-water Mixtures and Emulsions and Water Purification," *J. Colloid Interface Sci.* **557**, 395–407.

Liu, G. and Ding, J. (1998), "Diblock Thin Films with Densely Hexagonally Packed Nanochannels," *Adv. Mater.* **10**, 69–71.

Ma, W., Samal, S. K., Liu, Z., Xiong, R., De Smedt, S. C., Bhushan, B., Zhang, Q., Huang, C. (2017), "Dual pH- and Ammonia-vapor-responsive Electrospun Nanofibrous Membranes for Oil-water Separations," *J. Membr. Sci.* **537**, 128–139.

Negin, S., Daschbach, M. M., Kulikov, O. V., Rath, N., and Gokel, G. W. (2011), "Pore Formation in Phospholipid Bilayers by Branched-Chain Pyrogallol[4]arenes," *J. Am. Chem. Soc.* **133**, 3234–3237.

Ogasawara, W., Shenton, W., Davis, S. A., and Mann, S. (2000), "Template Mineralization of Ordered Macroporous Chitin–Silica Composites Using a Cuttlebone-Derived Organic Matrix," *Chem. Mater.* **12**, 2835–2837.

Peinemann, K.-V., Abetz, V., and Simon, P. F. W. (2007). "Asymmetric Superstructure Formed in a Block Copolymer via Phase Separation," *Nat. Mater.* **6**, 992–996.

Percec, V., Dulcey, A. E., Balagurusamy, V. S. K., Miura, Y., Smidrkal, J., Peterca, M., Nummelin, S., Edlund, U., Hudson, S. D., Heiney, P. A., Duan, H., Magonov, S. N., and Vinogradov, S. A. (2004), "Self-assembly of Amphiphilic Dendritic Dipeptides into Helical Pores," *Nature* **430**, 764–768.

Percec, V., Dulcey, A. E., Peterca, M., Adelman, P., Samant, R., Balagurusamy, V. S. K., and Heiney, P. A. (2007), "Helical Pores Self-Assembled from Homochiral Dendritic Dipeptides Based on l-Tyr and Nonpolar α-Amino Acids," *J. Am. Chem. Soc.* **129**, 5992–6002.

Phillip, W. A., Hillmyer, M. A., Cussler, E. L. (2010), "Cylinder Orientation Mechanism in Block Copolymer Thin Films Upon Solvent Evaporation," *Macromolecules* **43**, 7763–7770.

Pollard, S. J. T., Fowler, G. D., Sollars, C. J., Perry, R. (1992), "Low-cost Adsorbents for Waste and Waste-water Treatment–a Review," *Sci. Total Environ.* **116**, 31–52.

Sengur-Tasdemir, R., Aydin, S., Turken, T., Genceli, E. A., and Koyuncu, I. (2016), "Biomimetic Approaches for Membrane Technologies," *Sep. Purif. Rev.* **45**, 122–140.

Shannon, M. A., Bohn, P. W., Elimelech, M., Georgiadis, J. G., Mariñas, B. J., and Mayes, A. M. (2008), "Science and Technology for Water Purification in the Coming Decades," *Nature* **452**, 301–310.

Shin, Y., Liu, J., Chang, J. H., Nie, Z., and Exarhos, G. J. (2001), "Hierarchically Ordered Ceramics Through Surfactant-Templated Sol-Gel Mineralization of Biological Cellular Structures," *Adv. Mater.* **13**, 728–732.

Shin, Y., Wang, L.-Q., Chang, J. H., Samuels, W. D., and Exarhos, G. J. (2003), "Morphology Control of Hierarchically Ordered Ceramic Materials Prepared by Surfactant-directed Sol-gel Mineralization of Wood Cellular Structures," *Studies in Surface Science and Catalysis* **146**, 447–451.

Srivastava, A., Srivastava, O. N., Talapatra, S., Vajtai, R., and Ajayan, P. M. (2004), "Carbon Nanotube Filters," *Nat. Mater.* **3**, 610–614.

Surwade, S. P., Smirnov, S. N., Vlassiouk, I. V., Unocic, R. R., Veith, G. M., Dai, S., and Mahurin, S. M. (2015), "Water Desalination using Nanoporous Single-layer Graphene," *Nat. Nanotechnol.* **10**, 459–464.

Taguchi, A. and Schüth, F. (2005), "Ordered Mesoporous Materials in Catalysis," *Micropor. Mesopor. Mater.* **77**, 1–45.

Verkman, A. S., Anderson, M. O., and Papadopoulos, M. C. (2014), "Aquaporins: Important but Elusive Drug Targets," *Nat. Rev. Drug Discov.* **13**, 259–277.

Wang, S. and Peng, Y. (2010), "Natural Zeolites as Effective Adsorbents in Water and Wastewater Treatment," *Chem. Eng. J.* **156**, 11–24.

Yang, D., Qi, L., and Ma, J. (2002), "Eggshell Membrane Templating of Hierarchically Ordered Macroporous Networks Composed of TiO_2 Tubes," *Adv. Mater.* **14**, 1543–1546.

Zhang, B., Davis, S. A., and Mann, S. (2002), "Starch Gel Templating of Spongelike Macroporous Silicalite Monoliths and Mesoporous Films," *Chem. Mater.* **14**, 1369–1375.

Zhao, Y. Qiu, C., Li, X., Vararattanavech, A. Shen, W. Torres, J., Hélix-Nielsen, C., Wang, R., Hu, X., Fane, A. G., and Tang, C. Y. (2012), "Synthesis of Robust and High-performance Aquaporin-based Biomimetic Membranes by Interfacial Polymerization-membrane Preparation and RO Performance Characterization," *J. Membr. Sci.* **423–424**, 422–428.

Chapter 8
Selected Oil-Water Separation Techniques—Lessons from Living Nature

Species exist in living nature which either attract or repel water in air, and some are oil repellant underwater. By using nature's inspiration, porous bioinspired surfaces can be fabricated which provide a combination of water and oil repellency and affinity. These surfaces can be used for oil-water separation and water purification (Bhushan 2018, 2019a, b).

In this chapter, an overview of two species which exhibit superliquiphobicity/philicity is presented. For oil-water separation, some fabrication approaches for superliquiphobic/philic porous surfaces are presented.

8.1 Lotus Leaf and Shark Skin for Superliquiphobicity/philicy

Plant leaves found in living nature are either water repellant or have an affinity to water (Bhushan 2018). The lotus leaf is known to be super water repellent and is a model surface for superhydrophobicity (Koch et al. 2008, 2009; Bhushan and Jung 2011; Barthlott et al. 2017; Bhushan 2018). Lotus leaves are found in muddy ponds, and yet the leaf surface is typically clean. Water droplets falling on the leaf are found to exhibit high contact angles and a low contact angle hysteresis or tilt angle due to formation of air pockets (referred to as the Cassie-Baxter state of wetting). Therefore, droplets roll off the surface, moving easily across the leaf, collecting debris as they go and keeping the leaf clean for photosynthesis. The superhydrophobic nature of the leaf surface is a result of its chemistry and surface roughness. It is composed of a hierarchical structure of microbumps formed by convex papillae epidermal cells, covered with 3-D epicuticular wax which self-assemble in the form of nanotubules, Fig. 8.1a. The nanotubules are made of a hydrophobic wax and, when combined with the micropapillae, result in a superhydrophobic surface with low contact angle hysteresis or tilt angle (Bhushan 2018).

© Springer Nature Switzerland AG 2020
B. Bhushan, *Bioinspired Water Harvesting, Purification, and Oil-Water Separation*, Springer Series in Materials Science 299,
https://doi.org/10.1007/978-3-030-42132-8_8

Lotus leaf (*Nelumbo nucifera*)

Droplet on a lotus leaf surface

(a)

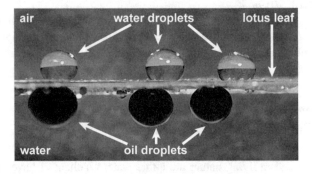

(b)

Fig. 8.1 **a** Surface morphology of a lotus leaf surface and a water droplet on top of the surface exhibiting high contact angles (adapted from Bhushan 2018) and **b** droplets of oil underwater exhibiting high contact angles when placed on the underside of the leaf (adapted from Cheng et al. 2011; Bhushan 2018)

Examples of oil repellency in nature are generally limited to underwater oil repellency (Bhushan 2018). As mentioned earlier, the top of the lotus leaf is superhydrophobic due to wax nanotubules. However, the underside of the leaf has no such structures and is superhydrophilic and superoleophilic. When floating on water, the underside of the leaf is superoleophobic, Fig. 8.1b. In another example, shark skin is superhydrophilic and superoleophilic in air but superoleophobic underwater, Fig. 8.2 (Bixler and Bhushan 2012, 2013; Bhushan 2018). Water soaks into the skin and forms a thin water layer. The oil droplets sit on top of the water

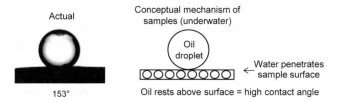

Fig. 8.2 Optical images of water droplet in air and oil droplet underwater for shark skin replicas (adapted from Bixler and Bhushan 2012, 2013)

layer to result in superoleophobicity. It should be noted that the surface is only superoleophobic underwater and that can be a major limitation in application.

8.2   Fabrication Approaches for Superliquiphobic/philic Porous Surfaces for Oil-Water Separation

Coated porous surfaces that attract one liquid (-philic) and repel the other liquid (-phobic) are suitable for oil-water separation. Oils have a lower surface tension than that of water (Brown and Bhushan 2016a; Bhushan 2018, 2019a, b). It is therefore possible to design surfaces that repel water but have an affinity for oils. Inspiration can be taken from the lotus leaf to create hierarchically structured surfaces with relatively low surface energy, lower than surface tension of water but higher than that of oil. The low surface energy will make the surface hydrophobic and the hierarchical roughness will result in a superhydrophobic surface with high water contact angle and low oil contact angle. Oil-water separation can be achieved when the coating is applied to a porous substrate (Bhushan 2019a, b). The oil will pass through the porous structure and water will be collected on the top surface.

Alternatively, a coating with an affinity to water and oil will make the surface with a combination of superhydrophilicity and superoleophobicity. In a coated porous surface, water will pass through the porous structure and oil will be collected on the top surface, providing oil-water separation (Bhushan 2019a, b).

Bioinspired roughness-induced surfaces which attract one liquid and repel the other liquid have been fabricated by various researchers (Brown and Bhushan 2015a, b; Bhushan 2018; Bhushan and Martin 2018; Chauhan et al. 2019; Li et al. 2019). To fabricate a roughness-induced superhydrophobic/superoleophilic surface, a non-fluorinated silane layer can be deposited on a surface with hierarchical roughness. Coated porous structures can be used for oil-water separation with oil passing through and water being collected on the top surface. To fabricate a superoleophobic/superhydrophilic surface, a fluorosurfactant layer can be used, which contains a high surface energy head group and a low surface energy tail group. After deposition, the fluorinated tails segregate at the air interface, resulting in a low surface energy barrier that repels oils. However, when droplets of water are placed on the surface, they are able to penetrate down through the tail groups to reach the high surface energy polar head groups below. The coating therefore appears hydrophilic while also being oleophobic. Therefore, porous structures coated with fluorosurfactants can be used for oil-water separation with water passing through and oil being collected on the top surface.

Re-entrant geometries are used for improved repellency of very low surface tension liquids. Re-entrant geometries are shapes that have overhanging structures in which the surface features become narrower at the base (Nosonovsky and Bhushan 2008; Brown and Bhushan 2016b; Bhushan 2018). These geometries are necessary for repelling low surface tension liquids, and particularly surfactant-containing liquids such as shampoos and laundry detergents due to their low tension components and active groups.

For superoleophilic porous surfaces, after absorbing a certain amount of oil, the separation efficiency of these surfaces reduces as the filter surface retains oil. In all superoleophilic porous surfaces, oil contamination is of concern, and porous materials must be cleaned or replaced are use. Therefore, superoleophobic/super-hydrophilic porous surfaces are preferred in which the water passes through the porous material and the oil is repelled. Additionally, water is denser than oil and tends to sink to the bottom of a mixture, meaning that hydrophobic/oleophilic materials are not suitable for certain applications, such as gravity-driven separation.

Porous structures commonly used include stainless steel mesh, fabric, filter paper, sponge, and compressible cotton (Bhushan 2019b). The stainless steel mesh can have pore size (opening) as low as about 25 μm. Woven cotton fabrics can also have pore size as low as about 25 μm (Angelova 2012). However, cotton can have very small pore size (on the order of fraction of 1 μm to about 10 μm) dependent upon compressibility. Different porous structures are used for separation of different oil-water mixtures. For immiscible oil-water mixtures, stainless steel mesh is commonly used. Cotton fabric is also occasionally used. For oil-water emulsions, the pore size of steel meshes and cotton fabrics is generally too large and compressible cotton is preferred. Cotton fabric can be used in some oil-water emulsions with relatively large liquid droplets of one liquid dispersed in the other. Photographs of various porous surfaces are shown in Fig. 8.3.

Unlike absorbent materials, in the coated porous material techniques, both phases are immediately separated with no additional steps required to remove one phase from the material. It is envisioned that such a device could be used upstream of the conventional purification membranes. This ensures that the majority of the contaminant phase is removed before further purification and processing, resulting in greater efficiency of the more selective membranes.

Porous surfaces

Immiscible oil-water mixture

Stainless steel mesh   Cotton fabric

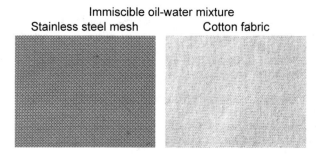

Oil- water emulsion
Cotton

Fig. 8.3 Photographs of stainless steel mesh, cotton fabric, and cotton, used as a porous medium for separating immiscible oil-water mixtures and oil-water emulsions (adapted from Bhushan 2019b)

References

Angelova, R. A. (2012), "Determination of the Pore Size of Woven Structures through Image Analysis," *Cent. Eur. J. Eng.* **2**, 129–135.

Barthlott, W., Mail, M., Bhushan, B., and Koch, K. (2017), "Plant Surfaces: Structures and Functions for Biomimetic Innovations," *Nano-Micro Letters* **9**, 23.

Bhushan, B. (2018), *Biomimetics: Bioinspired Hierarchical-Structured Surfaces for Green Science and Technology*, third ed., Springer International, Cham, Switzerland.

Bhushan, B. (2019a), "Lessons from Nature for Green Science and Technology: An Overview and Bioinspired Superliquiphobic/philic Surfaces," *Phil. Trans. R. Soc. A* **377**, 20180274.

Bhushan, B. (2019b), "Bioinspired Oil-water Separation Approaches for Oil Spill Clean-up and Water Purification," *Phil. Trans. R. Soc. A* **377**, 20190120.

Bhushan, B. and Jung, Y. C. (2011), "Natural and Biomimetic Artificial Surfaces for Superhydrophobicity, Self-cleaning, Low Adhesion, and Drag Reduction," *Prog. Mater. Sci.* **56**, 1–108.

Bhushan, B. and Martin, S. (2018), "Substrate-independent Superliquiphobic Coatings for Water, Oil, and Surfactant Repellency: An Overview," *J. Colloid Interface Sci.* **526**, 90–105.

Bixler, G. D. and Bhushan, B. (2012), "Bioinspired Rice Leaf and Butterfly Wing Surface Structures Combining Shark Skin and Lotus Effects," *Soft Matter* **8**, 11271–11284.

Bixler, G. D. and Bhushan, B. (2013), "Fluid Drag Reduction and Efficient Self-cleaning with Rice Leaf and Butterfly Wing Bioinspired Surfaces," *Nanoscale* **5**, 7685–7710.

Brown, P. S. and Bhushan, B. (2015a), "Mechanically Durable, Superoleophobic Coatings Prepared by Layer-by-layer Technique for Anti-smudge and Oil-water Separation," *Sci. Rep.–Nature* **5**, 8701.

Brown, P. S. and Bhushan, B. (2015b), "Bioinspired, Roughness-induced, Water and Oil Super-philic and Super-phobic Coatings Prepared by Adaptable Layer-by-layer Technique," *Sci. Rep.-Nature* **5**, 14030.

Brown, P. S. and Bhushan, B. (2016a), "Bioinspired Materials for Water Supply and Management: Water Collection, Water Purification and Separation of Water from Oil," *Phil. Trans. R. Soc. A* **374**, 20160135.

Brown, P. S. and Bhushan, B. (2016b), "Designing Bioinspired Superoleophobic Surfaces," *APL Mater.* **4**, 015703.

Chauhan, P., Kumar, A., and Bhushan, B. (2019), "Self-cleaning, Stain-resistant and Anti-bacterial Superhydrophobic Cotton Fabric Prepared by Simple Immersion Technique," *J. Colloid Interface Sci.* **535**, 66–74.

Cheng, Q., Li, M., Zheng, Y., Su, B., Wang, S. and Jiang, L. (2011), "Janus Interface Materials: Superhydrophobic Air/solid Interface and Superoleophobic Water/solid Interface Inspired by a Lotus Leaf," *Soft Matter* **7**, 5948–5951.

Koch, K., Bhushan, B., and Barthlott, W. (2008), "Diversity of Structure, Morphology, and Wetting of Plant Surfaces," *Soft Matter* **4**, 1943–1963.

Koch, K., Bhushan, B., and Barthlott, W. (2009), "Multifunctional Surface Structures of Plants: An Inspiration for Biomimetics," *Prog. Mater. Sci.* **54**, 137–178.

Li, F., Bhushan, B., Pan, Y., and Zhao, X. (2019), "Bioinspired Superoleophobic/superhydrophilic Functionalized Cotton for Efficient Separation of Immiscible Oil-Water Mixtures and Oil-Water Emulsions," *J. Colloid Interface Sci.* **548**, 123–130.

Nosonovsky, M. and Bhushan, B. (2008), *Multiscale Dissipative Mechanisms and Hierarchical Surfaces: Friction, Superhydrophobicity, and Biomimetics*, Springer, Heidelberg, Germany.

Chapter 9
Bioinspired Oil-Water Separation and Water Purification Approaches Using Superliquiphobic/philic Porous Surfaces and External Stimuli

In water contaminated with oil, oil and water are not soluble or miscible. Oil-water mixtures present can be divided into immiscible mixtures and emulsions, Fig. 9.1 (Bhushan 2019b). Immiscible oil-water mixtures are common in water contamination and oil spills. Oil-water emulsions are common in some liquid waste. In the immiscible oil-water mixture, mixtures are somewhat stratified by densities of oil and water. An emulsion is a type of colloid formed by combining two liquids that normally do not mix. It is a more intensive mix type in which microdroplets (less than a micron to a few microns) of one liquid are dispersed in the other. Emulsions normally require mechanical agitation and the use of emulsifiers (surface-active agents). Emulsifiers are used to prevent the suspended droplets from coalescing and breaking up the emulsion. Based on dispersed phases, oil-water emulsions fall into two types: oil-in-water and water-in-oil emulsions (Kokal 2005; Prince 2012). An example of an oil-in-water emulsion would be milk and water-based paint. An example of water-in-oil emulsion would be margarine and mayonnaise.

Oil-water separation techniques are of interest in industrial applications and important for environmental sustainability. To separate immiscible oil-water mixtures, traditional physical methods, such as gravity separation, centrifuge, sedimentation and hydrocyclone separation, are widely used (Rao et al. 2012; Prince 2012; Bhushan 2018). These methods are either time-consuming or energy intensive. For the more complex emulsions, thermo/chemical demulsifiers and electrolytic demulsification methods are commonly used. However, the complexity and the high energy consumption add to the cost.

For oil spill remediation, various methods are used by the oil industry (Brown and Bhushan 2016; Bhushan 2018). Toxic dispersants are used in which chemicals reduce the interfacial tension between oil and water to facilitate the breakup of the oil into smaller droplets. However, the toxicity and poor biodegradability of both dispersants and the resulting dispersed oil makes using such chemicals undesirable. Skimming of the oil from the surface of the water with absorbent booms is another method of oil removal, however this is dependent upon favorable conditions including calm waters and slow oil speeds. Other methods for the collection of the

© Springer Nature Switzerland AG 2020
B. Bhushan, *Bioinspired Water Harvesting, Purification,
and Oil-Water Separation*, Springer Series in Materials Science 299,
https://doi.org/10.1007/978-3-030-42132-8_9

Immiscible oil-water mixtures   Emulsions

                                        Oil-in-water   Water-in-oil

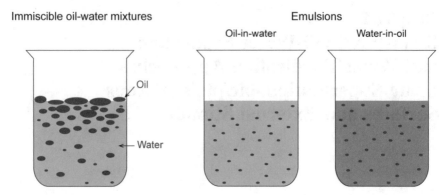

Fig. 9.1 Schematics of immiscible oil-water mixtures and miscible oil-in-water and water-in-oil emulsions (adapted from Bhushan 2019b)

oil use other absorbent materials such as zeolites and organoclays, or natural fibers such as straw, cellulose, wool, or human hair. However, many of these materials also have a tendency to absorb water, which can lower their efficiency. Additionally, the absorbed oil must be removed from the material, making such methods incompatible with continuous flow systems (Brown and Bhushan 2016). An introduction to various oil-water separation techniques commercially used for oil spill cleanup is presented in Appendix 9.A.

For water purification, separation filters and membranes that repel one liquid phase while allowing the other to pass through also exist. These are commonly designed for maximum permeation, at the cost of some degree of selectivity.

Bioinspired superliquiphobic/philic porous surfaces may provide optimum solutions for oil-water separation and water purification, which are sustainable and environmentally friendly. In this chapter, a review of various bioinspired approaches for fabrication of superliquiphobic/philic porous surfaces with affinity to water and repellency to oil or vice versa is presented. Details of fabrication and characterization of superliquiphobic/philic stainless steel mesh, cotton fabrics, and cotton are presented. Applications of bioinspired oil-water separation techniques to oil spill cleanup and water purification are discussed. Finally, water purification technique using ultraviolet (UV) light stimulus is presented.

9.1 Coated Stainless Steel Mesh for Separation of Immiscible Oil-Water Mixtures

Superhydrophobic/superoleophilic and Superoleophobic/superhydrophilic stainless steel mesh surfaces having multi-scale roughness with low surface energy coatings have been developed. In an early example of fabrication of superhydrophobic/ superoleophobic mesh surfaces, only a low surface energy coating was used to repel water. A polytetrafluoroethylene (PTFE) emulsion was spray coated onto stainless steel meshes with various pore diameters (Feng et al. 2004). The resulting coated mesh was superhydrophobic and superoleophilic with contact angles of about 150° for water and about 0° for diesel oil. When oil was added to the mesh, it quickly spread and permeated the mesh and was collected on the other side, while water added to the mesh remained on top.

Later, bioinspired approaches were developed for fabrication of superhydrophobic/superoleophobic mesh surfaces. Brown and Bhushan (2015a, b) used a nanoparticle/binder layer to create hierarchical roughness by using a so-called, layer-by-layer technique. The nanoparticles also increase the coating hardness, resulting in a durable coating. Two polyelectrolyte layers were used to support a silica nanoparticle layer by electrostatic attraction on a stainless steel mesh surface. They deposited a silane layer on the top to produce a surface which was superhydrophobic and superoleophilic. Bhushan and Martin (2018) used a spray coating of silica nanoparticles and methyphenyl silicone binder to produce a low surface energy surface with hierarchical roughness. Both coated mesh surfaces demonstrated that hexadecane oil goes through the mesh and water is collected on the top. Both deposition techniques are facile.

Nanda et al. (2019) developed another facile method. They chemically etched steel mesh to create flower like microstructures, then coated the etched surface with hexadecyltrimethoxysilane (HDTMS) to produce a superhydrophobic and superoleophilic surface. They demonstrated that kerosene-water and n-hexane-water mixtures can be separated with kerosene or n-hexane going through the filter with a high separation efficiency.

In an early example of fabrication of underwater superoleophobic surfaces, these were created by immersing stainless steel mesh into an acrylamide solution. The acrylamide was polymerized by UV irradiation to form a hydrogel, a water-swollen polymer network (Xue et al. 2011). When placed underwater, the treated hydrophilic mesh exhibited oil contact angles of 153° and droplets were found to roll easily from the tilted surface. When an oil-water mixture was poured onto the mesh, the water permeated through the mesh while the oil remained on top. Although this configuration helps to reduce the impact of oil fouling, the fact that the surface is only superoleophobic underwater limits its application.

Switchable superhydrophobic/superoleophilic and superhydrophilic/underwater superoleophobic stainless steel mesh and cotton surfaces have been produced by using photoinduced superliquiphobicity/philicity of TiO_2 (Li et al. 2019b). They developed switchable superliquiphobic/superliquiphilic coating by modifying TiO_2 particles with 1H, 1H, 2H, 2H-perfluorodecyltrimethoxysilane (PFDMS). They used UV stimulus and heating to achieve switchability for oil-water separation applications.

Brown and Bhushan (2015b) created a superoleophobic/superhydrophilic mesh surface by combining the chemistry of the fluorosurfactant with hierarchical roughness using the layer-by-layer technique described earlier. Bhushan and Martin (2018) used the nanoparticle/binder coating with an over layer of fluorosurfactant. The fluorosurfactant provides oil repellency and water affinity, while the roughness enhances these properties to result in oil contact angles of >150° and water contact angles of <5°. When this coating is applied to a stainless steel mesh and an oil–water mixture is poured over it, the water penetrates through the mesh while the oil remains on top and can be easily rolled off by tilting. By inclining the mesh at an angle, the two liquids can be separated and collected simultaneously. A nanoparticle/binder composite coating technique is found to work with a variety of substrates with desired performance (Bhushan and Martin 2018). It requires a minimum number of processing steps, which is desirable (Martin et al. 2017).

Such superoleophobic/superhydrophilic surfaces represent the ideal scenario for oil-water separators (Bhushan 2019a, b). Their oil-repellent nature means they are less prone to oil fouling than devices where the water phase is being repelled. In addition, their water affinity makes them more compatible with gravity-driven separation, where the water phase will end up at the bottom of the mixture, or in scenarios where water is the dominant phase.

Fabrication techniques for superliquiphobic/philic stainless steel meshes and selected characterization data and their application for oil-water separation will be presented next.

9.1.1 Fabrication Technique

Schematic of fabrication techniques to produce superhydrophobic/superoelophilic and superoleophobic/superhydrophilic coatings are shown in Fig. 9.2 (Bhushan 2019b). Coatings consisting of hydrophobic SiO_2 nanoparticles and a binder of methylphenyl silicone resin with or without functional layers to obtain combinations of superhydrophobic/philic and superoleophobic/philic properties have been fabricated (Bhushan and Martin 2018; Bhushan 2018, 2019a). Methylphenyl silicone resin binder is commonly selected because it is known to be durable and offers strong adhesion between the nanoparticles and substrate. Hydrophobic, 10 nm SiO_2 nanoparticles are selected because they have high hardness for wear resistance and high visible transmittance for transparency. For a superhydrophobic/superoleophobic surface, the nanoparticle/binder mixture is deposited by a spray method on various substrates. For superoleophobic/superhydrophilic surfaces, the nanoparticle/binder coating is irradiated with ultraviolet-ozone (UVO) treatment followed by spin/spray deposition of fluorosurfactant (aqueous anionic fluorosurfactant, Capstone FS-50).

For the coating mixture, 375 mg of hydrophobic silica nanoparticles (10 nm diameter, Aerosil RX300, Evonik Industries) were dispersed in 30 mL of 40% tetrahydrofuran (THF, Fisher Scientific) and 60% IPA by volume. This mixture was sonicated using an ultrasonic homogenizer (20 kHz frequency at 35% amplitude, Branson Sonifer 450A) for 15 min. Then, 150 mg of methylphenyl silicone resin (SR355S, Momentive Performance Materials) was added for a particle-to-binder ratio (p-b ratio) of 2.5 for optimum wettability with highest value of CA and lowest value of TA. The mixture was then sonicated for an additional 15 min to form the final mixture. One mL of the coating mixture was deposited via spray gun

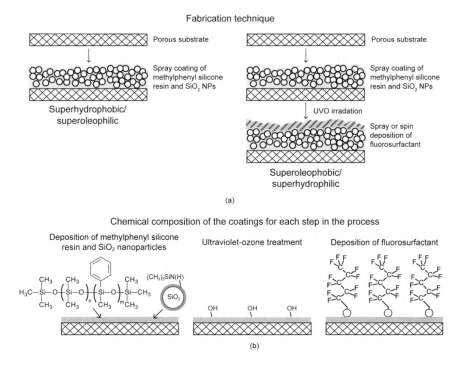

Fig. 9.2 Schematics of fabrication techniques. For superhydrophobic and superoleophilic surfaces, hydrophobic nanoparticles and methylphenyl silicone binder is applied to provide nanoroughness. For superhydrophobic and superoleophilic surfaces, nanoparticle/binder coating is treated by ultraviolet-ozone light to activate the surface and then fluorosurfactant is deposited as the last step (adapted from Bhushan 2019b)

(Paasche®) from 10 cm away with compressed air at 210 kPa. The sample was transferred to an oven at 70 °C for 5 min to remove the remaining solvent. The coated surfaces were superhydrophobic/superoleophilic (Bhushan and Martin 2018; Bhushan 2018, 2019a, b).

For superoleophobic/superhydrophilic surfaces, a fluorosurfactant coating was applied on top of the nanoparticle/binder coating. To prepare an active surface, the underlayer was first irradiated using UVO treatment with the samples placed 2 cm underneath the lamp source for 60 min. The UVO exposure was generated from a U-shaped, ozone producing, ultraviolent lamp (18.4 W, Model G18T5VH-U, Atlantic Ultraviolet Co.). It is reported that this lamp puts out a total of 5.8 W of 254 nm light, 0.4 W of 185 nm light, and 1.6 g/h of ozone in ambient conditions. The lamp was placed in an enclosure and was connected to an electronic ballast (120 v, Model 10-0137, Atlantic Ultraviolet Co.) in order to provide the proper electrical conditions (Bhushan and Martin 2018). Next, 1 mL of a fluorosurfactant solution (Capstone FS-50, DuPont) diluted with ethanol to an overall fluorosurfactant concentration of 45 mg/mL was spin coated or spray coated onto the sample (Bhushan and Martin 2018; Bhushan 2018, 2019a, b).

9.1.2   Characterization of Coated Glass Surfaces

Characterization data of nanoparticle/binder coatings with and without fluorosurfactant on glass substrates is presented. The glass substrate is a commonly used substrate in scientific studies.

9.1.2.1   Surface Morphology

To study surface morphology, scanning electron microscope (SEM) and atomic force microscope (AFM) images of a superoleophobic coating on glass substrates were taken, Fig. 9.3 (Bhushan and Martin 2018). SEM images show the multi-scale roughness structure of the coating. From the AFM measurements, root mean square (RMS) roughness of the coating was about 1 µm with a Peak-to-Valley (P-V) distance of 5.5 µm.

The coating thickness was obtained by measuring the step height of the coating deposited on part of the substrate. The measured coating thickness at particle-to-binder (p-b) ratio of 2.5 was about 3 µm.

Surface morphology of superoleophobic coating

Top view SEM images at two magnifications

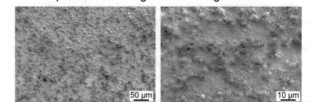

AFM surface height map
RMS = 1.0±0.1 µm, P-V = 5.5±0.2 µm

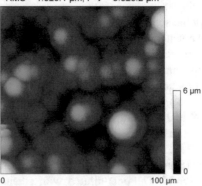

Fig. 9.3 SEM and AFM images of superoleophobic coating with an optimum p-b ratio of 2.5 on a glass substrate (adapted from Bhushan and Martin 2018)

9.1.2.2   Wettability

Contact angles and tilt angles were measured using a standard automated goniometer. Optical images and measured contact angle (CA) and tilt angle (TA) values for water and hexadecane droplets on the glass substrate and with two coatings are shown in Fig. 9.4 (Bhushan and Martin 2018). Untreated glass has a water CA of $55 \pm 2°$ and a hexadecane CA of $27 \pm 1°$. By applying the coatings to these substrates, super-hydrophobic/superoleophilic and superoleophobic/superhydrophilic properties were obtained. The nanoparticle/binder coating resulted in water CA of $165 \pm 2°$, water TA $\leq 1°$, and wetting with hexadecane. The nanoparticle/binder and fluorosurfactant coating results in wetting with water, hexadecane CA of $157 \pm 2°$, and hexadecane TA of $2 \pm 1°$.

9.1.2.3   Wear Resistance

Wear resistance on the macroscale of the superoleophobic coating on glass substrate was investigated by performing wear tests using a ball-on-flat tribometer (Bhushan 2013a, b). For the wear test, a 3 mm diameter sapphire ball was slid

Fig. 9.4 Optical images and measured CA and TA values of water and hexadecane droplets on glass substrates and coated samples to show repellency and wetting with the two liquids (adapted from Bhushan and Martin 2018)

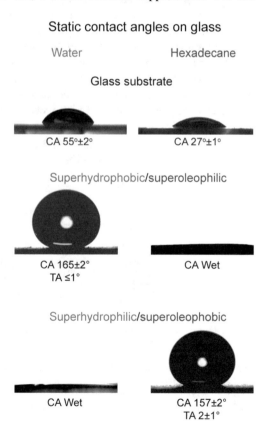

Static contact angles on glass

Water   Hexadecane

Glass substrate

CA 55°±2°   CA 27°±1°

Superhydrophobic/superoleophilic

CA 165±2°   CA Wet
TA ≤1°

Superhydrophilic/superoleophobic

CA Wet   CA 157±2°
                                            TA 2±1°

Wear experiment on superoleophobic surfaces using ball-on-flat tribometer

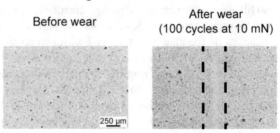

Fig. 9.5 Optical micrographs after wear experiment using ball-on-flat tribometer using a 3 mm diameter sapphire ball with 10 mN load on superoleophobic coating. There was slight burnishing but the coating was still able to repel hexadecane (adapted from Bhushan and Martin 2018)

against the samples in a reciprocating mode. The test was carried out at 10 mN for 100 cycles. The optical images of the samples after wear experiments showing the wear track are shown in Fig. 9.5 (Bhushan and Martin 2018). Burnishing of the coatings was observed. The coating was still able to repel hexadecane over the burnished region suggesting that the coating was not destroyed.

To study wear resistance on the microscale, the superoleophobic coating was scanned with an AFM using a 15 μm radius borosilicate glass ball mounted on a rectangular lever at a normal load of 10 μN over 50 μm × 50 μm region (Bhushan 2017). Figure 9.6 shows AFM images over 100 μm × 100 μm scan area before and after the AFM wear experiments. After the AFM test, there was no burnishing of the coating. There was no discernible difference between the surface morphologies of regions before and after wear (Bhushan and Martin 2018).

Wear experiment using AFM on superoleophobic surface

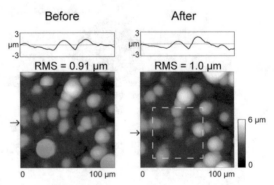

Fig. 9.6 AFM surface height maps and surface profiles (locations indicated by arrows) before and after AFM wear experiments with 15 μm radius borosilicate glass ball at a load of 10 μN over 50 μm × 50 μm on superoleophobic coating. After AFM test, there was no burnishing of the coating (adapted from Bhushan and Martin 2018)

9.1.3   Characterization of Coated Stainless Steel Mesh Surfaces for Oil-Water Separation

Stainless steel meshes (#400) were cleaned with acetone and 2-propanol (Fisher Scientific) until they were found to be hydrophilic. They were then coated with nanoparticle/binder coating and some with fluorosurfactant coating as well to produce superhydrophobic/superoleophilic and superoleophobic/superhydrophilic meshes, respectively.

To study repellency to oil and affinity to water droplets on a superoleophobic/superhydrophilic stainless steel mesh, droplets were placed on the coated mesh, as shown in Fig. 9.7. Oil droplets remained on the mesh. The sequence of photographs of the water droplet on the coated mesh shows that the water droplet spreads quickly and penetrates through the mesh in a fraction of a second (Bhushan 2019b).

To demonstrate oil-water separation capability in continuous flow systems, agitated, immiscible oil-water mixtures were poured onto coated stainless steel meshes suspended horizontally over beakers as shown in Fig. 9.8. In both cases, the -philic component quickly passed through the mesh, while the -phobic component remained on top of the mesh. When the mesh was tilted, the -phobic component rolled across the top of the mesh and was collected in another beaker (Brown and Bhushan 2015a, b; Bhushan and Martin 2018; Bhushan 2018).

As indicated earlier, the use of oleophobic/hydrophilic coated surfaces is preferable to using hydrophobic/oleophilic coated surfaces because surface contamination by oil and other oil-based contaminants is common, and the porous material must then be cleaned or replaced, resulting in a drop in the separation efficiency.

9.1.4   Applications to Oil Spill Cleanup and Water Purification

Oil-water separation is of interest in oil spill cleanup and water purification. Methods commonly used for oil spill cleanup include the use of toxic dispersants to break up oil globs into smaller droplets, controlled burning which is not eco-friendly, solvents to soak up oil on contact (very slow method), and skimmer and booms. An environmentally friendly and fast approach would be to use coated steel nets that are superoleophobic/superhydrophilic which should allow oil-water separation in a continuous flow (Brown and Bhushan 2016). Figure 9.9 shows a schematic of how these bioinspired nets can be incorporated into currently used

Oil and water droplets on superoleophobic/ superhydrophilic stainless steel mesh

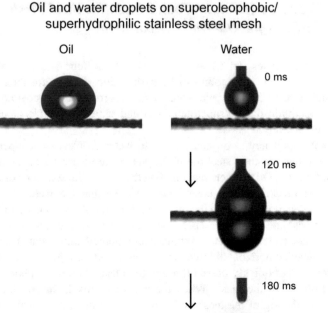

Fig. 9.7 Photographs of oil and water droplets on superoleophobic/superhydrohilic stainless steel mesh. Oil droplet remained on the mesh surface, whereas, water droplet spread quickly and penetrated through the mesh in a fraction of a second (adapted from Bhushan 2019b)

booms on a boat (Brown and Bhushan 2016). When the boat is driven in the oil spill region, the bioinspired nets will capture the oil, where it can be recovered by pumping, while allowing water to pass through. These nets can be used repeatedly.

The bioinspired nets can also be used for water purification by removing organic contamination in water.

Photograph of oil-water separation
Mesh placed on an inclined plane

Superhydrophilic/Superoleophobic

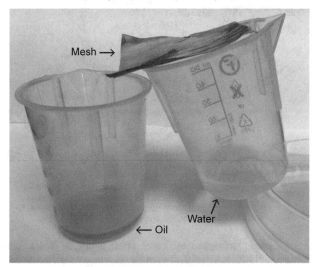

Superhydrophobic/Superoleophilic

Fig. 9.8 Photographs of the superoleophobic/superhydrophilic and superhydrophobic/super-oleophilic coated stainless steel meshes acting as oil-water separators. On the superoleophobic/superhydrophilic coated mesh, oil collects on top of the mesh while water passes through. In contrast, on the superhydrophobic/superoleophilic coated mesh, oil passes through the mesh while the water remains on the top surface. If the meshes are placed at an angle, oil or water sitting on the mesh can be collected simultaneously in separate beakers. Red oil and blue water dyes were used to enhance contrast (adapted from Brown and Bhushan 2015b; Bhushan and Martin 2018)

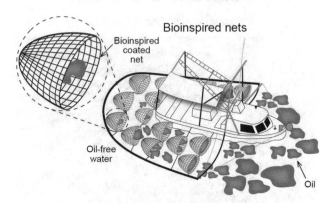

Schematic showing bioinspired nets connected to a boat for oil-spill clean up

Bioinspired nets

Bioinspired coated net

Oil-free water

Oil

Nets featuring bioinspired coated mesh trap oils
while allowing passage of water. Oil can easily
be recovered from nets via pumping. Oil-repellent
nature of mesh reduces need for cleaning.

Fig. 9.9 Schematic showing bioinspired nets connected to a boat for oil spill cleanup. When to boat is driven in an oil spill region, nets will capture the oil, where it can be recovered by pumping, while allowing water to pass through (adapted from Bhushan 2018)

9.1.5 Summary

For fabrication of superliquiphobic/philic surfaces, the nanoparticle/binder coating method is attractive due to its ease of deposition, flexibility in substrate application, and desirable properties. The nanoparticle/binder coating method can be used for fabrication of superhydrophobic/superoleophobic and superoleophobic/superhydrophilic steel meshes for oil-water separation and water purification.

9.2 Coated Cotton Fabric for Separation of Immiscible Oil-Water Mixtures

In addition to stainless steel mesh, cotton fabric can be used for immiscible oil-water separation. Superhydrophobicity of cotton fabric is achieved by introducing multi-scale roughness and coating with low a surface energy material (Bhushan 2018). Treatment of the cotton fabric surfaces is challenging due to the inherent heterogeneous roughness with low thermal stability. Several methods such as dip coating (Zeng et al. 2016; Zhu et al. 2017), solution immersion coating (Panda et al. 2018; Chauhan et al. 2019; Tudu et al. 2019), sol-gel based coating (Przybylak et al. 2016), wet chemical process (Wang et al. 2017), and spray coating (Hsieh et al. 2011) have been used to make cotton fabric superhydrophobic. Physical and chemical durability has been a concern. Chauhan et al. (2019) reported that wettability was maintained after physical and chemical abuse and demonstrated high durability. Chauhan et al. (2019) fabricated superhydrophobic cotton fabric by simple immersion in non-fluorinated hexadecyltrimethoxysilane (HDTMS) solution. Their coating repelled oils with surface tension as low as 47.70 mN m^{-1} (ethylene glycol). Tudu et al. (2019) fabricated superhydrophobic cotton fabric by simple immersion in perfluorodecyltriethoxysilane (PFDTS) solution. Their coating repelled oils with surface tension as low as 27.05 (hexadecane). Both Chauhan et al. (2019) and Tudu et al. (2019) demonstrated that coated fabric can be used for oil-water separation as long as surface tension of oil in the mixture was low enough that it wetted the coated surface.

A facile fabrication technique and selected characterization data for superhydrophobic/superoleophilic cotton fabric and their applications for oil-water separation will be presented next (Chauhan et al. 2019).

9.2.1 Fabrication and Characterization Techniques

To fabricate superhydrophobic cotton fabric, as received treated cotton fabric was cleaned ultrasonically to remove wax and other impurities with distilled water and ethanol for 30 min and subsequently dried at 70 °C for 1 h. A 5% (v/v) HDTMS was dissolved in ethanol and stirred for 1 h at room temperature to obtain a homogeneous solution. Subsequently, cleaned cotton fabric was immersed in HDTMS solution for 5 h and then dried at 120 °C for 6 h in a hot air oven to remove the solvent. Finally, it was left in air for drying (Chauhan et al. 2019).

Surface morphologies of as received and treated cotton fabric were examined with SEM. The wettability of the treated cotton fabric was studied by measuring contact angles with a 3–5 µL droplet of water and liquids (with surface tension greater than 47 mN/m) such as tea, honey, milk, and ethylene glycol. Surface tension values of selected liquids are given in Table 9.1. Chemical stability of the treated cotton fabric was examined by immersing samples in saline water (3.5% w/v NaCl) and organic solvents (chloroform, toluene, and dimethyl carbonate). At regular intervals of immersion, contact angles were measured to observe any changes in wettability. A thermal stability test was performed by heating the samples at various temperatures (120, 140, 160, 180, and 200 °C) for 1 h. After cooling the samples, the contact angles were measured (Chauhan et al. 2019).

To evaluate the self-cleaning performance, as received and treated cotton fabric were immersed in muddy water for 10 min. The treated cotton fabric was difficult to immerse in the muddy water due to the high liquid repellence property, so it was immersed by application of force. Afterward, the effect of mud on the surfaces was characterized by photographing samples before and after soaking in mud (Chauhan et al. 2019).

Table 9.1 Surface tension values of various liquids and whether superhydrophobic/super-oleophilic cotton repels liquid or not (adapted from Chauhan et al. 2019)

Liquid	Surface tension (mN m^{-1})	Treated cotton repels or not
Water	71.99[a]	Repel
Glycerol	63.40[b]	Repel
Milk	55–60[c]	Repel
Ethylene glycol	47.70[a]	Repel
Kerosene	~ 30[d]	Does not repel
Benzene	28.80[e]	Does not repel
n-Hexane	18.43[a]	Does not repel

[a]Rumble (2018)
[b]Takamura et al. (2012)
[c]Chandan (1997)
[d]Speight (2017)
[e]Harkins and Brown (1919)

For oil-water separation studies, before conducting an experiment, the treated cotton fabric was dipped in oil (organic solvents). The wetted treated cotton fabric was placed on top of a funnel. Later, the oil-water mixture was poured onto treated cotton fabric which repelled the water (blue color) and was collected on its surface and the oil penetrated the fabric by gravity. Next, n-hexane, kerosene, and ethylene glycol were used as oils in the oil-water mixtures. Separation efficiency was calculated by dividing volumes of water after the separation process by volumes of water before the separation process (Chauhan et al. 2019).

9.2.2   Surface Morphology and Wettability

The surface morphologies of as received and treated cotton fabric were observed by using SEM as shown in Fig. 9.10 (Chauhan et al. 2019). The SEM images show no significant differences due to the formation of HDTMS monolayers, however, both surfaces show the hierarchical morphology.

Wettability of as received and treated cotton fabric was examined by measuring contact angles. Water contact angles and tilt angles are presented in Fig. 9.10. It was found that as received cotton fabric exhibits superhydrophilic and super-oleophilic behavior with a contact angle of $\sim 0°$. After modification of cotton fabric with HDTMS, it displays water repellency with a water contact angle of $157 \pm 2°$. A water droplet easily rolled off with a tilt angle of $7 \pm 1°$, indicating its super-hydrophobic and self-cleaning nature (Chauhan et al. 2019).

Images of various liquid droplets on treated cotton fabric are shown in Fig. 9.11 and observations on whether the fabric repels or not selected liquids along with their surface tension values are presented in Table 9.1. Liquids whose surface tension is equal to or more than 47 mN/m (tea, colored water, milk, lemon,

As received cotton fabric

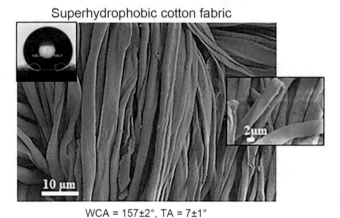

WCA = 0°

Superhydrophobic cotton fabric

WCA = 157±2°, TA = 7±1°

Fig. 9.10 SEM images of as received and superhydrophobic/superoleophilic cotton fabric. Water contact angle and tilt angles are also presented (adapted from Chauhan et al. 2019)

turmeric, honey, glycerol and ethylene glycol) show good repellency for the treated cotton fabric with contact angles more than 150°, revealing the superoleophobic nature (Chauhan et al. 2019).

9.2.3   Physical and Chemical Durability

Treated cotton fabrics should work indoors as well as in harsh environments such as high temperature, UV irradiation, acidic/alkaline/organic contact, and washing. The durability after washing was evaluated by immersing superhydrophobic cotton fabric into a detergent solution, organic solvent (benzene) and hot water (80 °C), and then it was ultra-sonicated for 1 h. Afterward wettability behavior of the dried

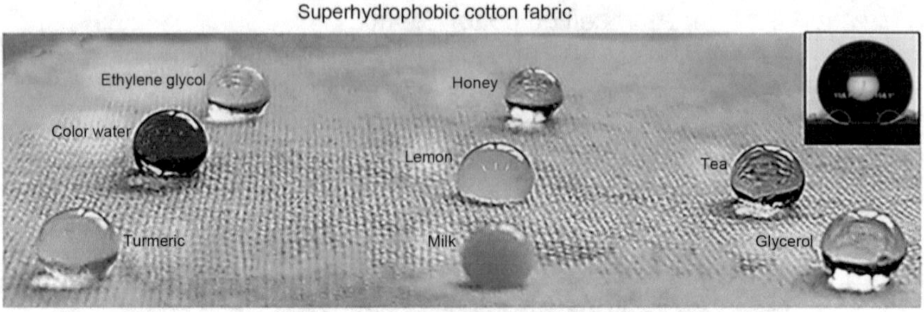

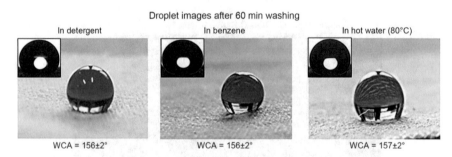

Fig. 9.11 Optical images of various droplets (color water, tea, milk, lemon, turmeric, honey, glycerol and ethylene glycol) on superhydrophobic cotton fabric. Liquid repellency for all liquid droplets is observed (adapted from Chauhan et al. 2019)

Fig. 9.12 Optical images of water droplets on treated cotton fabric after 60 min washing in detergent solution, benzene and hot water (80 °C) via ultra-sonication method, showing its excellent washing ability. Inserts are contact angle images. Water contact angles are also presented (adapted from Chauhan et al. 2019)

fabric was examined by measuring the contact angle. Data are shown in Fig. 9.12 (Chauhan et al. 2019). It was observed that contact angles of all washed fabrics remained at more than 150° and water droplets rolled off the surface, exhibiting the washable nature of the surface without any damage. This test indicates that the HDTMS-treated cotton fabric maintained wettability properties after washing.

The chemical stability of the treated cotton fabric was examined by immersing the samples in organic solvents (chloroform, toluene, and dimethyl carbonate) for 7 days and in saline water (3.5% w/v NaCl) for 24 h. After immersion, water droplets on the surface formed spherical shapes as shown in Fig. 9.13 (Chauhan et al. 2019). Contact angles were found to remain more than 150°, displaying superhydrophobic nature. Thus, treated cotton fabric was suitable for organic solvents, as well as the saline water, which is important for industrial purposes.

The thermal stability of the treated cotton fabric was studied by heating the sample at various temperatures (120–200 °C) for 1 h in a hot air oven. The data are shown in Fig. 9.14 (Chauhan et al. 2019). The superhydrophobicity, with water contact angle of 157 ± 2° of the treated cotton fabric, was maintained after heating

After 7 days in chloroform

WCA = 156±2°

After 7 days in toluene

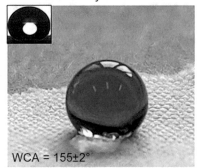

WCA = 155±2°

After 7 days in dimethyl carbonate

WCA = 156±2°

After 12 h in saline water

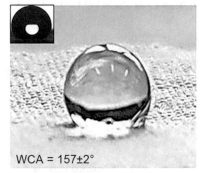

WCA = 157±2°

Fig. 9.13 Optical images of water droplets on treated cotton fabric after immersion in chloroform for 7 days, toluene for 7 days, dimethyl carbonate for 7 days, and saline water for 12 h. Inserts are contact angle images. Water contact angles are also presented. Chemical stability is demonstrated (adapted from Chauhan et al. 2019)

Fig. 9.14 Contact angle of treated cotton fabric after heating for 1 h at different temperatures. Thermal stability of treated cotton fabric is demonstrated below 150 °C (adapted from Chauhan et al. 2019)

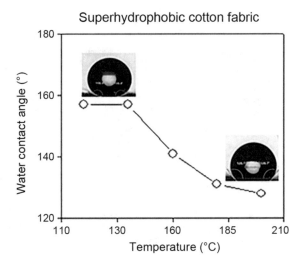

up to 120–150 °C, exhibiting thermal stability of the treated cotton fabric below 150 °C. The contact angle of the cotton was 141° after heating to 160 °C and the superhydrophobicity turned into hydrophobicity due to the decomposition of the HDTMS (boiling point, 155 °C) which resulted in the removal of low surface energy material from the surface (Chauhan et al. 2019).

9.2.4 Self-cleaning Properties

The self-cleaning properties of the treated cotton fabric were examined by immersing as received and treated cotton fabric into muddy water for 10 min, as shown in Fig. 9.15 (Chauhan et al. 2019). It was observed that the treated cotton fabric did not immerse easily in mud and remained floating on the muddy water

Fig. 9.15 Optical images of as received and superhydrophobic cotton fabric before and after immersion in the muddy water for 10 min, demonstrating self-cleaning property of treated cotton fabric (adapted from Chauhan et al. 2019)

surface due to its high water repellence property. Therefore, it was forcefully immersed into the muddy water. After immersion, as received cotton fabric was found to be wetted with muddy water and it remained polluted even after cleaning and drying. On the other hand, treated cotton fabric was fully clean when it was removed from the muddy solution, showing its excellent self-cleaning property (Chauhan et al. 2019).

9.2.5   Separation of Immiscible Oil-Water Mixtures

Droplets of water and liquids with a surface tension more than 47 mN/m could not penetrate through the superhydrophobic cotton fabric surface whereas low surface tension oils could. Therefore, coated cotton fabric can be used for oil-water separation with oils having surface tension equal to or less than 47 mN/m. Oil (n-hexane = 18.43 mN/ m), kerosene ($\sim$30 mN/m) and ethylene glycol (47.70 mN/m) were used to separate

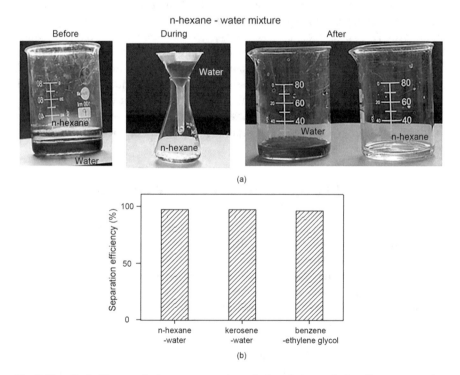

Fig. 9.16 a Optical images of n-hexane-water mixture before, during, and after oil-water separation and separation efficiency of n-hexane-water, kerosene-water mixture, and benzene-ethylene glycol mixtures, by using treated cotton fabric. Water was dyed blue color to visualize the difference between oil and water mixture. **b** Separation efficiency with three mixtures was about 99% (adapted from Chauhan et al. 2019)

from their mixtures. All mixtures consisted of a 1:1 ratio. Figure 9.16a shows optical images of n-hexane-water mixture before, during, and after oil water separation (Chauhan et al. 2019). It can be seen that oil is separated by its own gravity. After separation, no water trace (blue color) is visible within the penetrated oil (n-hexane), showing excellent oil-water separation efficiency. Separation efficiency of oil-water (n-hexane-water and kerosene-water) and oil liquid (benzene-ethylene glycol) mixture was measured and is presented in Fig. 9.16b. It was found to be about 99%. After a few cycles, the separation process became slow due to adsorption of the oil on the cotton surface. Afterward, oil adsorbed cotton fabric was cleaned with acetone and water. It was found that the cleaned cotton fabric showed a contact angle more than 150° and could be reused for the oil-water separation process (Chauhan et al. 2019).

9.2.6   Summary

Superhydrophobic cotton fabric was produced by using a simple immersion technique in non-fluorinated HDTMS solution (Chauhan et al. 2019). Treated cotton fabric exhibited repellency with water and liquids with surface tension equal to or more than 47 mN/m, such as tea, honey, glycerol, ethylene glycol and milk. Wettability was maintained after machine washing in detergent solution, benzene and hot water at 80 °C, after immersion in chloroform, toluene, and dimethyl carbonate, and in saline water, as well as exposure to high temperature (~ 150 °C). The treated cotton exhibited self-cleaning properties. Treated cotton fabric could separate the oil from its oil-water (n-hexane-water, kerosene-water) and oil-liquid (benzene-ethylene glycol) with high separation efficiency of about 99%.

9.3   Coated Cotton for Separation of Oil-Water Emulsions

The mesh-based materials with opposite wettability towards oil and water are used in separating immiscible oil-water mixtures. Separation efficiency is dependent upon the pore size. The mesh-based materials are impractical for separating emulsions, because the microdroplets dispersed in emulsions can easily pass through the mesh whose apertures are larger than about 25 µm. To separate emulsions, one of the effective approaches is to decrease the pore size of the substrates. Attempts have been made to accomplish the separation of emulsions by selecting specially-made mesh or membrane substrates whose pore sizes are several micrometers (Gao et al. 2014; Zeng et al. 2016). The efforts of decreasing the pore sizes resulted in some success in emulsion separation, but were not efficient for separation of immiscible oil-water mixtures with high flux.

To separate both immiscible oil-water mixtures and emulsions, a compressible soft material which has small pore structures would be desirable. The use of cotton makes it possible to separate immiscible and emulsified mixtures. A superhydrophobic/superoleophilic 3D porous structure with fine pore size was used to separate oil from immiscible oil-water mixtures as well as from water-in-oil emulsions (Liu et al. 2019). However, a porous structure of this kind is easily contaminated and the pores become blocked by oil. Li et al. (2019a) developed superoleophobic/superhydrophilic cotton for oil-water separation. The coated cotton provided oil-repellency and water-wetting behavior in ambient atmosphere. It was capable of separating both immiscible oil-water mixtures and oil-in-water emulsions. Since the cotton pores do not get blocked by oil, this approach is attractive.

A facile fabrication technique and selected characterization data for superoleophobic/superhydrophilic cotton and their applications for oil-in-water separation will be presented next (Li et al. 2019a).

9.3.1 Fabrication and Characterization Techniques

9.3.1.1 Fabrication Technique

The fabrication process is shown in Fig. 9.17a (Li et al. 2019a). Fluorosurfactant and hydrophilic Al_2O_3 nanoparticles were mixed in an ethanol solution with a magnetic stirrer and functionalized nanoparticles were formed by the attachment between the fluorosurfactant and Al_2O_3 particles in the process. Next, the cotton fibers were coated with the functionalized nanoparticles. The cotton fibers get attached to nanoparticles and fluorinated chains with low surface energy arranged at the air interface, as shown in Fig. 9.17b.

To determine an optimum concentration of fluorosurfactant, wetting properties were measured as a function of various concentrations. It was found that there was no effect on the superoleophobicity and superhydrophilicity when the concentration of the fluorosurfactan in ethanol solution was higher than 30 mg mL^{-1}. To prepare a fluorosurfactant-ethanol solution with a concentration of 40 mg mL^{-1}, 2 g fluorosurfactant (Capstone^{TM} FS-50, Dupont) was dissolved in a 50 mL ethanol solution followed by 30 min of magnetic stirring. To determine the optimal concentration of Al_2O_3 nanoparticles, the superoleophobicity (contact angles and tilt angles of hexadecane droplets) of treated cotton was investigated by varying the Al_2O_3 concentration from 20 to 100 mg mL^{-1}. They found that the contact angle of hexadecane remained around 150° when the concentration was higher than 20 mg mL^{-1}, whereas, the tilt angle was fairly high when the concentration was higher than 80 mg mL^{-1} or lower than 40 mg mL^{-1}. The lowest tilt angle occurred at the concentration of about 60 mg mL^{-1}. Therefore, 3 g Al_2O_3 nanoparticles (30 nm, Aladdin Bio-chem) were added into the aforementioned solution so that the Al_2O_3 concentration was 60 mg mL^{-1}. The solution was magnetic stirred for another 20 min to improve mixing to produce a paint-like suspension (Li et al. 2019a).

Next, a piece of pristine cotton ($\sim$0.5 g, Aladdin Bio-chem) was initially washed with ethanol and deionized water in succession, followed by 40 min of

Fabrication technique

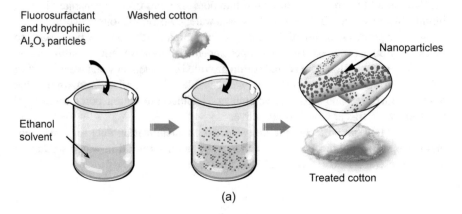

(a)

Chemical attachment of fluorosurfactant to cotton

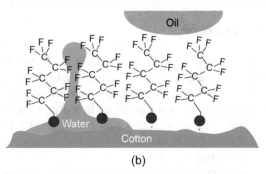

(b)

Fig. 9.17 **a** The fabrication technique of the superoleophobic/superhydrophilic cotton, and **b** chemical attachment of fluorosurfactant to cotton (adapted from Li et al. 2019a)

vacuum drying at 60 °C. The cotton then was immersed in the suspension and ultrasonically cleaned for 20 min before being dried at atmospheric pressure at 80 °C (Li et al. 2019a).

9.3.1.2   Preparation of Emulsions

To prepare emulsions, diesel fuel (surface tension ~ 25.05 mN m^{-1}) and hexade-cane oil (surface tension ~ 27.05 mN m^{-1}) were selected as the dispersed phase, and TWEEN60 (MW 1131, Aladdin Bio-chem) was selected as an emulsifier. TWEEN60, fuel or oil and deionized water were mixed in a ratio of 1:10:100 by volume. Each solution was mixed with an ultrasonic homogenizer for 3 h. All emulsions appeared to be stable for more than 20 h under atmospheric conditions (Li et al. 2019a).

9.3.1.3   Wettability

Wettability was determined by measuring contact angles and tilt angles of the liquid droplets on the treated cotton. The droplet volume used was about 5 μL. Selected liquids of scientific and engineering interest were deionized water, diesel fuel, and octane, hexadecane, and dichloroethane oils. Their surface tension data is presented in Table 9.2.

9.3.1.4   Separation Method of Immiscible Mixtures and Emulsions

To separate immiscible oil-water mixtures by a gravity-driven method, the treated cotton and a stainless steel mesh (#300) used as a support was pressed together between two glass tubes. The gap between the tubes was then sealed by silicone sealant (Silco RVT-4500). The combination was then set above the beaker with the cotton side up. For gravity driven separation, an immiscible oil-water mixture was poured into the top tube and the water penetrated the treated cotton and fell into the beaker (Li et al. 2019a).

For separation of emulsions, superoleophobic/superhydrophilic cotton (~ 1 g) was pushed into an injector barrel as tightly as possible and the density of the compressed cotton was measured to be about 0.28 g cm^{-3}. An oil-in-water emulsion was poured into the barrel, as shown in Fig. 9.18 (Li et al. 2019a). During the separation, water will wet and penetrate the compressed cotton, while the microdroplets of oil will be rejected.

A 500 ml of emulsion feed was used for measurement of flux and separation efficiency of oil-in-water emulsions. The oil content in the feed and filtrate was measured by a total organic carbon analyzer. The dispersion images of the emulsions were taken and the separated flux was calculated by dividing the volume of the separated emulsion divided by the product of the sectional area of the injector barrel and the time of the separation process. The efficiency of the emulsion-separation process was calculated by dividing the difference between concentrations of oil in the emulsion feed and filtrate, divided by the concentration of the oil in the emulsion feed (Li et al. 2019a).

Table 9.2 Surface tensions, contact angles, and tilt angles of liquids on superoleophobic/superhydrophilic cotton (adapted from Li et al. 2019a)

Liquid	Surface tension (mN m^{-1})	Contact angle (°)	Tilt angle (0°)
Deionized water	71.99[a]	Wetted	Wetted
1, 2-Dichloroethane	31.86[a]	153 ± 2	4 ± 1
Hexadecane	27.05[a]	154 ± 2	4.5 ± 1
Diesel	25–29	153 ± 2	5.5 ± 1
Octane	21.14[a]	152 ± 2	4 ± 1

[a]Rumble (2018)

Fig. 9.18 The schematic of the injector barrel for separation of emulsions. The treated cotton is pushed into the injector barrel and compressed as tightly as possible (density is about 0.28 g cm^{-3}), then the emulsions is poured into the barrel (adapted from Li et al. 2019a)

Injector barrel for separation of emulsions

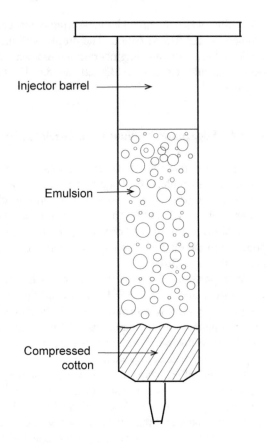

9.3.2 Surface Morphology and Wettability

The surface morphology of the cotton was characterized by SEM and the images are shown in Fig. 9.19 (Li et al. 2019a). The fibers in the treated cotton were covered by the agglomerated fluorinated Al$_2$O$_3$ nanoparticles. Overlapping distributed fibers as well as the nanoparticles constitute a multi-dimensional roughness in both macro and micro scales. This multi-scale structure plays a dominant role in providing superoleophobicity and superhydrophilicity (Bhushan 2018). Surface roughness of the pristine and treated cotton was measured by using laser scanning confocal microscopy. The root mean square roughness, R$_q$, is presented in Fig. 9.19. Treated cotton is rougher than the pristine cotton.

To assess the superoleophobicity and superhydrophilicity of the treated cotton, contact angles and tilt angles of fuel, oil and water were also measured. Data are summarized in Table 9.2 (Li et al. 2019a). Images of droplets of various liquids on the treated cotton are shown in Fig. 9.20a (Li et al. 2019a). Diesel fuel and various

Pristine cotton (R_q = 0.51 μm)

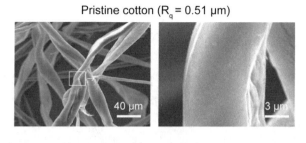

Treated cotton (R_q = 1.65 μm)

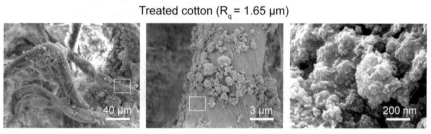

Fig. 9.19 SEM images of pristine and treated cotton with the roughness data (adapted from Li et al. 2019a)

oils with different surface tensions were repelled by the treated cotton without sticking or wetting. On the other hand, water droplets wetted and permeated the cotton upon immediate contact with the fibers. The hexadecane oil droplets were even able to slide freely on the treated cotton with low contact angle hysteresis, which demonstrated the oil repellent behavior of the treated cotton as shown in Fig. 9.20b (Li et al. 2019a).

9.3.3   Separation of Oil-Water Mixtures

9.3.3.1   Immiscible Oil-Water Mixtures

The treated cotton could separate many immiscible oil-water mixtures by a simple gravity-driven pouring method. Figure 9.21 shows that the water, dyed blue, continuously passed through the cotton and hexadecane, dyed red, was restrained on the surface without any penetration. This treated cotton is expected to be superior to superhydrophobic/superoleophilic materials during which the material is easily blocked by viscous oil.

9.3.3.2   Oil-in-Water Emulsions

The superoleophobic/superhydrophilic cotton could also be used for the separation of oil-in-water emulsions due to its superoleophobicity and superhydrophilicity. A schematic of separation of oil-in-water emulsion is shown in Fig. 9.22 (Li et al.

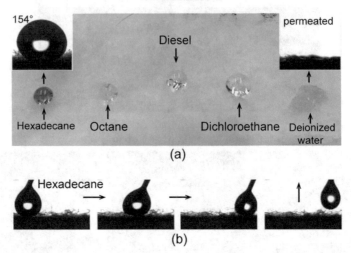

(a)

(b)

Fig. 9.20 a Optical images of water and oil droplets on treated cotton showing superoleopho-bicity and superhydrophilicity of the treated cotton. **b** Successive images of the hexadecane droplet sliding on the treated cotton. A droplet of hexadecane was pushed on the cotton and was slid from left to right before detachment (adapted from Li et al. 2019a)

Fig. 9.21 Gravity-driven separation of immiscible oil-water mixture (adapted from Li et al. 2019a)

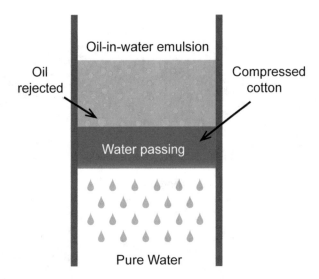

Fig. 9.22 Schematic of separation of oil-in-water emulsion. During the separation, the water phase wetted and permeated the compressed cotton, and the dispersed oil droplets were rejected (adapted from Li et al. 2019a)

2019a). For separation of oil-in-water emulsions, the continuous water phase easily wets the superhydrophilic fibers and permeates the cotton, whereas, the continuous water phase as well as the wetted fibers on the top surface form an obstructive layer towards oil and thus the micro oil droplets were rejected at the surface of the compressed cotton. Since the oil droplets do not stick on the cotton surface, the oil should not block the pores of the compressed cotton.

To test the separation properties, optical micrographs of the emulsion feed and filtrate, before and after separation were taken, as shown in Fig. 9.23 (Li et al. 2019a). Diesel-in-water and hexadecone-in-water emulsion s were used as emulsion feeds. The figure shows that oil droplets were randomly dispersed in both emulsion feeds, while there were no droplets detected in the filtrates after separation.

To quantify the separation properties, the separated flux and the separation efficiency were measured, as shown in Fig. 9.24 (Li et al. 2019a). High separated flux was observed which varied from 520 to 650 L m^{-2} h^{-1}. The separation efficiency was higher than 98% for continuous separation of both diesel-in-water and hexadecane-in-water separation. High efficiency provides a practical approach in low-cost and effective separation of different stabilized emulsions in both industrial and domestic areas.

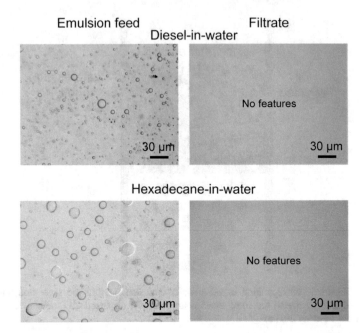

Fig. 9.23 Optical images of the emulsion feeds and filtrates before and after separation. Diesel-in-water and hexadecane-in-water emulsions were used as emulsion feeds

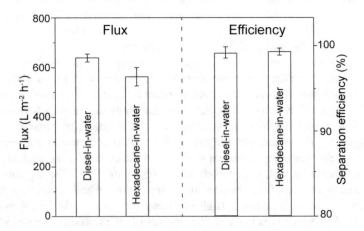

Fig. 9.24 The separated flux and separation efficiency for diesel-in-water and hexadecane-in-water emulsions (adapted from Li et al. 2019a)

9.3.4 Summary

A superoleophobic/superhydrophilic cotton was developed by immersion in a solution of fluorosurfactant and hydrophilic alumina particles in ethanol (Li et al. 2019a). The treated cotton, which exhibited oil-repellent and water-wetting behavior, can be used in separating various immiscible oil-water mixtures without oil contamination. In addition, the treated cotton can also be used for separating oil-in-water emulsions with high flux and high efficiency without oil contamination.

9.4 TiO$_2$-Based Material Using UV Stimulus for Water Purification

Materials with switchable wettability have been fabricated by taking advantage of the photoinduced property of TiO$_2$ based materials. TiO$_2$ nanoparticles can be made superhydrophobic by bonding them with fluorinated materials and producing a low surface energy coating with hierarchical structure. During UV irradiation, photoinduction is stimulated and hydroxyl groups (–OH) are generated on the surface of TiO$_2$ particles. The hydroxyl groups have strong attraction with water which leads to the superhydrophilicity (and underwater-superoleophobicity). By heating the coating, the hydroxyl groups are replaced by oxygen in air which leads to conversion from superhydrophilicity to superhydrophobicity. These materials with reversible wettability switching between superhydrophobicity/superoleophilicity and superhydrophilicity/underwater-superoleophobicity, can be used for various oil-water separation or filtration applications (Liu et al. 2014; Li et al. 2019b).

In addition to photoinduced properties, photocatalysis is another intrinsic property of TiO$_2$. Both photoinduced and photocatalysis properties can take place simultaneously on the surface of TiO$_2$ under UV irradiation, with different mechanisms. For the photocatalysis process, highly oxidizing radicals are generated on the TiO$_2$ surface during UV irradiation. These radicals, which can subsequently degrade contaminants, have applications in liquid purification (Ohama and Van Gemert 2011). Therefore, TiO$_2$ based materials which include both photoinduced and photocatalysis properties have found application in liquid purification.

Li et al. (2019b) developed multi-functional, switchable superliquiphobic/liquiphilic coatings by simply modifying TiO$_2$ nanoparticles with 1H,1H,2H,2H-perfluorodecyltrimethoxysilane (PFDMS). The coating can be sprayed or dip coated on various substrates and provides the superhydrophobic/superoleophilic property. The coating can be rapidly switched between superhydrophobicity/superoleophilicity and superhydrophilicity/underwater-superoleophobicity by alternating the UV irradiation and heating processes. During UV irradiation, the coating can also degrade contamination in water due to the photocatalytic property of TiO$_2$. Fabrication technique and selected characterization data for water purification will be presented next.

9.4.1 Fabrication Technique of a Switchable Superliquiphobic/philic Coating

Schematic of fabrication steps are shown in Fig. 9.25 (Li et al. 2019b). For fabrication of the coating, 1.2 g of 1H,1H,2H,2H-perfluorodecyltrimethoxysilane (PFDMS, 98%, Aladdin Bio-chem) was dissolved in 50 mL ethanol and followed by a 30 min magnet stirring at ambient temperature. Then TiO_2 nanoparticles (P25, Degussa) with a TiO_2/PFDMS mass ratio of 4 were added into the solution and stirred for another 30 min to form a homogeneous suspension. The suspension can be sprayed or dip coated on rigid substrates. The coated surface was then dried at room temperature for 10 min.

Inspired by lotus leaf, low surface energy and hierarchical roughness are two key factors in fabrication of superhydrophobic surfaces (Bhushan 2018). In this study, TiO_2 particles were bonded by PFDMS, the fluorinated chains in PFDMS provide low surface energy, while the TiO_2 particles in the coating offer a hierarchical structure to provide superhydrophobicity. The surface morphology of the coating is shown in Fig. 9.26 (Li et al. 2019b). The hierarchical roughness is observed.

During UV irradiation, photoinduction is stimulated and hydroxyl groups (–OH) will be generated on the surface of TiO_2 particles, Fig. 9.25. The hydroxyl groups have a strong attraction with water which leads to the superhydrophilicity. During the heating process, the hydroxyl groups are replaced by oxygen in the air which lead to a reconversion from superhydrophilicity to superhydrophobicity (Li et al. 2019b).

For switching of the wettability from superhydrophobicity to superhydrophilicity, a 30 W UV lamp (254 nm wavelength) was used as the stimulating source. The coated substrate was irradiated by the lamp with a UV power density of 94 $\mu W\ cm^{-2}$ and 5 cm above the substrate. In order to check the change of wettability, the water contact angle was measured every 2.5 min during irradiation. For

Fabrication steps of a TiO_2-based coating with switchable wettability for water purification

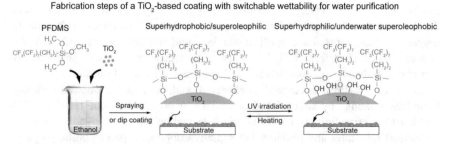

Fig. 9.25 Schematic of fabrication steps of a TiO_2-based coating with switchable wettability for water purification. TiO_2 nanoparticles are coated by 1H,1H,2H,2H-perfluorodecyltrimethoxysilane (PFDMS) which provides the low surface energy for superhydrophobicity. For conversion to superhydrophilicity, the hydroxyl groups (–OH) are generated on TiO_2 during UV irradiation which lead to superhydrophilicity/underwater-superoleophobicity. The hydroxyl groups can be removed by a heating process and the coating switches back to superhydrophobicity/superoleophilicity (adapted from Li et al. 2019b)

Fig. 9.26 Surface morphology of the TiO$_2$/PFDMS coating. The coating has a multiscale roughness (adapted from Li et al. 2019b)

SEM morphology of TiO$_2$/PFDMS coating

reverse switching of wettability, the coated substrate was heated at 150 °C in air environment for 80 min (Li et al. 2019b).

9.4.2 Photocatalytic Degradation of Contaminants for Water Purification

Mechanism of photocatalytic degradation of contaminants is shown in Fig. 9.27 (Li et al. 2019b). The photocatalytic property of TiO$_2$ nanoparticles was induced by ultraviolet light whose energy is larger than the band gap (energy difference between the top of the valence band (VB) and the bottom of the conduction band (CB)). This resulted in the generation of charge carriers (e$^-$ and h$^+$). The electrons (e$^-$) react with the oxygen in air to generate the superoxide radicals (O$_2^{\bullet-}$), while the holes (h$^+$) can react with water to generate hydroxyl radicals ($^\bullet$OH). Both O$_2^{\bullet-}$ and $^\bullet$OH have strong oxidizing properties and thereby result in the degradation of water-soluble contamination (Ohama and Van Gemert 2011). UV irradiation would also make the surface superhydrophilic. If the substrate is a porous surface, purified water will pass through the superhydrophilic mesh and can be collected underneath. The coated surface could then be reheated to realize the superhydrophobic property for reuse (Li et al. 2019b).

9.4.3 Water Purification Studies

Methylene blue (MB), which is an organic compound often used as a target for photocatalytic evaluation, was selected as water-soluble contamination. 20 mL of MB aqueous solution with a concentration of 2 ppm was poured onto the mesh. The

Mechanism of photocatalytic degradation of contaminants for water purification

Superhydrophobic/superoleophilic   Superhydrophilic/underwater superoleophobic

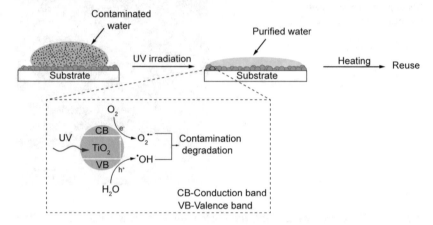

Fig. 9.27 Mechanism of photocatalytic degradation of contaminants siting on TiO_2-based coating for water purification. The photocatalytic property of TiO_2 nanoparticles can be induced by UV irradiation whose energy is larger than the band gap [the energy difference between the top of the valence band (VB) and the bottom of the conduction band (CB)]. Hydroxyl radicals ($^\bullet OH$) and superoxide radicals ($O_2^{\bullet -}$) are generated during this process, both of which have a strong oxidizing property and result into the degradation of water-soluble contamination. The coating could be reheated to realize the superhydrophobic property for reuse (adapted from Li et al. 2019b)

test solution was analyzed by UV spectrophotometer every 4 min to assess the efficiency of water purification. The purification experiment could be repeated by reheating the mesh to 150 °C in air environment for 80 min (Li et al. 2019b).

The water purification experiment was carried out using a superhydrophobic/superoleophilic coated stainless steel mesh (mesh number 300). The schematic of the water purification process is shown in Fig. 9.28a (Li et al. 2019b). Before UV irradiation, the contaminated water gets rejected on coated mesh due to the superhydrophobic property. During UV irradiation, soluble contamination in water is gradually degraded. Meanwhile, the coated mesh switched to superhydrophilic behavior and the purified water penetrated through the mesh for collection (Fig. 9.28b). To continue to use the coated mesh for the purification experiment, it could be heated to switch back to superhydrophobicity. The contaminated water which is rejected on the mesh before UV irradiation is visibly blue, whereas purified water which collected in the beaker after UV irradiation is transparent and clear.

To assess the efficiency of water purification, UV spectra of the methylene blue aqueous solution was measured at different times during the irradiation process, as shown in Fig. 9.29 (Li et al. 2019b). The peak which corresponded to MB experienced an obvious drop during the purification process, demonstrating the degradation of contamination.

Water purification process

Superhydrophobic/superoleophilic stainless steel mesh

(a)

(b)

Fig. 9.28 Schematic of water purification process showing purification of methylene blue aqueous solution on superhydrophobic/superoleophobic stainless steel mesh. **a** Photocatalytic degradation of contaminants occurs by UV irradiation. Pure water passes through the mesh. The mesh could be reused after heating it up. **b** The methylene blue in water was degraded during UV irradiation and transparent purified water flowed down into a beaker (adapted from Li et al. 2019b)

The durability of the coated mesh for water purification was studied by repeated irradiation-heating experiments. The coated mesh was reheated for recovery of superhydrophilicity after each purification. As shown in Fig. 9.30a, the coated mesh still maintains a hierarchal rough structure even after six purification cycles, which demonstrates the durability of the coating (Li et al. 2019b). The UV spectra of the test aqueous solution at different times during the irradiation process during the sixth cycle of the purification experiment, is shown in Fig. 9.30b. A continuous drop of MB is shown in the sixth cycle similar to that in Fig. 9.29. The data demonstrates the reusability of the coating for water purification.

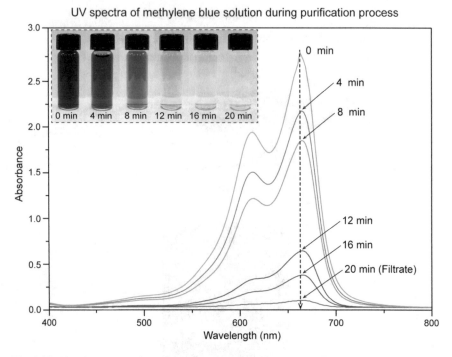

Fig. 9.29 The UV spectra of methylene blue aqueous solutions at different times of photodegradation showing the decrease of methylene blue concentration. The inset pictures are the methylene blue solutions after photodegradation at various times (adapted from Li et al. 2019b)

9.4.4 Summary

A UV-driven switchable superliquiphobic/liquiphilic TiO_2 based coating has been developed by a simple one-pot method. The coating can rapidly switch the wettability between superhydrophobicity/superoleophilicity and superhydrophilicity/underwater-superoleophobicity by UV irradiation and heating process. Owing to the photocatalysis property of TiO_2, the coated mesh also exhibits ability in on-line degradation of soluble contaminants in water under UV irradiation for water purification. For reuse of the coated mesh for water purification, switchable wettability can be used (Li et al. 2019b).

9.5 Closure

Oil contamination can occur during operation of machinery, oil exploration and transportation, and due to operating environment. Water contamination with various chemicals is another major concern with growing population and unsafe industrial

Mesh morphology before purification

Mesh morphology after six purificaiton cycles

(a)

UV spectra of methylene blue solution during the sixth cycle of purification process

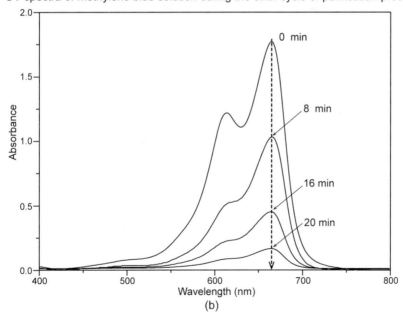

(b)

Fig. 9.30 **a** The morphology of coated mesh before and after six cycles of water purification. **b** UV spectra of methylene blue aqueous solutions at different times of photodegradation during the sixth cycle of purification (adapted from Li et al. 2019b)

practices of waste disposal. Commonly used oil-water separation techniques are either time consuming, energy intensive, and/or environmentally unfriendly.

Bioinspired superhydrophobic/superoleophilic and superoleophobic/superhydrophilic porous surfaces have been developed for oil-water separation which are sustainable and environmentally friendly. Stainless steel mesh and cotton fabric as porous materials are used for immiscible oil-water mixtures, and cotton for oil-water emulsions. In superoleophilic/superhydrophobic porous surfaces, oil contamination is of concern, and porous materials must be cleaned or replaced. Therefore, superoleophobic/superhydrophilic porous surfaces are preferred. Furthermore, for separation of oil-in-water emulsions, only superhydrophobic/superoleophilic porous surfaces can be used. Whereas, for separation of water-in-oil emulsions, only superoleophobic/superhydrophilic porous surfaces can be used.

Table 9.3 presents a summary of various filtration techniques for immiscible oil-water mixtures and emulsions with some comments. These filtration techniques can be used for various applications including oil spill cleanup, emulsion separation, and water purification.

Table 9.3 Summary of filtration techniques suitable for immiscible oil-water mixtures and oil-in-water and water-in-oil emulsions (adapted from Bhushan 2019b)

Filtration technique				
Mixtures to be filtered	Porous material and its wettability			
	Stainless steel mesh/cotton fabric		Cotton	
	Superhydrophobic/ superoleophilic	Superoleophobic/ superhydrophilic	Superhydrophobic/ superoleophilic	Superoleophobic/ superhydrophilic
Immiscible oil-water mixture	X	X	X	X
Oil-in-water emulsion	Pore size may be too large for effective separation		Continuous phase (water) does not penetrate and dispersed phase (oil) remains at the top as well	X Continuous phase (water) penetrates and oil gets rejected as it travels through the thickness of the porous material
Water-in-oil emulsion			X Continuous phase (oil) penetrates and water gets rejected as it travels through the thickness of the porous material	Continuous phase (oil) does not penetrate and dispersed phase (water) remains at the top as well

Filtration in immiscible mixture is dependent upon liquid surface tensions. Filtration in emulsion is dependent upon continuous phase and dispersed phase

TiO$_2$-based materials using UV light stimulus have also been developed for water purification.

Appendix 9.A: Introduction to Various Oil-Water Separation Techniques Commercially Used for Oil Spill Cleanup

Contamination of ground water and ocean water is of concern. Various industrial accidents on the ground and offshore occur periodically which lead to water contamination. Table 9.A.1 presents a list of notable offshore oil spills (Brown and Bhushan 2016). The Deepwater Horizon in the Gulf of Mexico in 2010 was the largest offshore oil spill in history with some 206 million gallons of oil being released in the Gulf. In addition, the emergence of fracking in the US, where water-based fluids (containing sand and chemicals) are injected under high pressure to fracture rocks for the release of previously inaccessible oil and gas, has led to a large increase in domestic oil production, Fig. 9.A.1 (Brown and Bhushan 2016). However, the process has also led to an increase in the amount of oil-contaminated wastewater. It is estimated that, between 2005 and 2014, some 248 billion US gallons of water were used for shale gas and oil extraction in the United States (Kondash and Vengosh 2015). In addition to this large amount of wastewater that must be processed before release, there is the potential for accidental contamination of ground and surface water.

Oil contamination occurs as part of routine operation in various industrial machinery. Examples include industrial lubricating oil effluents in electric transformers, air brakes of commercial vehicle systems, industrial compressors, and various components in mining and paper manufacturing industries. Oil contaminants need to be captured in some applications to maintain efficient operation of

Table 9.A.1 List of notable offshore oil spills (adapted from Brown and Bhushan 2016)

Spill	Location	Year	Amount spilled (million US gallons)
Deepwater Horizon[a]	Gulf of Mexico	2010	206
Gulf War[b]	Persian Gulf	1991	240
ABT Summer[c]	Angola	1991	80
MT Haven[c]	Italy	1991	42
Valdez[c]	Alaska	1989	11
Odyssey[c]	Canada	1988	43
Nowruz[b]	Persian Gulf	1983	80
Castillo de Bellver[c]	South Africa	1983	79
Ixtoc I[b]	Gulf of Mexico	1979	140
Atlantic Empress[c]	West Indies	1979	88
Amoco Cadiz[c]	France	1978	69

[a]Ramseur (2015)
[b]Etkin and Welch (1997)
[c]ITOPF (2015)

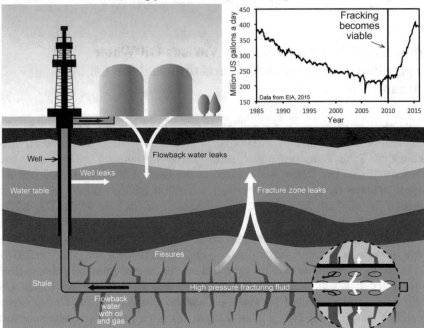

Fig. 9.A.1 Schematic of the fracking process. High pressure fluid (typically water, sand, and chemicals) is pumped underground, creating fissures and releasing previously inaccessible pockets of oil and gas, which are then removed along with the fluid. Leaks of contaminated fracking fluid into groundwater are possible from the fracture zone, the well, and flowback water storage containers. Inset: US crude oil production. Increase in production from 2010 is due to emergence of fracking, primarily occurring onshore in the lower 48 states (adapted from Brown and Bhushan 2016)

machinery and avoid environmental impact. Separating of oil from industrial wastewater is of interest to allow water re-use.

Finally, removal of oil and other organic contaminants from water for human consumption is critical. This is particularly important in the parts of the world with high contamination levels and limited water resources.

Oil spills cleanup and separation of oil from water for industrial re-use and human consumption remains an important environmental challenge. Various methods are commercially used for oil spill cleanup (Dave and Ghaly 2011; Fingas 2011; Brown and Bhushan 2016).

9.A.1 Dispersants

Dispersants are chemical compounds that reduce the interfacial tension between oil and water to promote the breakup of oil globs into smaller droplets. Dispersants are toxic and harmful to marine life, as well as expensive, and they decrease biodegradability of the oil (Nwaizuzu et al. 2015). Dispersants are sometimes are used to prevent oil from getting to the shore, however, oil that has been treated with dispersants that does get to shore will sink deep into the ground where it cannot easily be degraded or captured. Lastly, dispersants can cause oil to sink where it is both harder to capture, and slower to degrade due to the lower temperature.

9.A.2 Controlled Burning

Controlled burning is another method of disposing of oil. It is quick and effective; however, burning release a large amount of greenhouse gases as well as toxic compounds. Burning must be done far away from population centers to prevent health effects of the toxic gases released. To burn effectively, there must be at least 3 mm of oil on the surface of water.

9.A.3 Sorbents

Sorbents or sponges are materials that will soak up oil on contact, Fig. 9.A.2. Absorbent materials include zeolites, organoclays, or natural fibers such as straw, cellulose, wool, or human hair. The absorbent materials often are non-selective, absorbing both oil and water, lowering their efficiency (Zhu et al. 2011; Bayat et al. 2015). They also can sink when saturated with oil, making it difficult to recover. After, sucking up oil, sorbents would either have to be thrown away or cleaned, making them incompatible with continuous processes. This method is very slow and is not suitable for cleanup of large surface area.

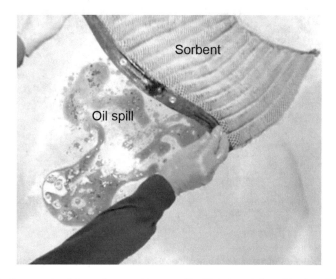

Fig. 9.A.2 Photograph of the sorbent sucking up oil spill (adapted from Bhushan 2018)

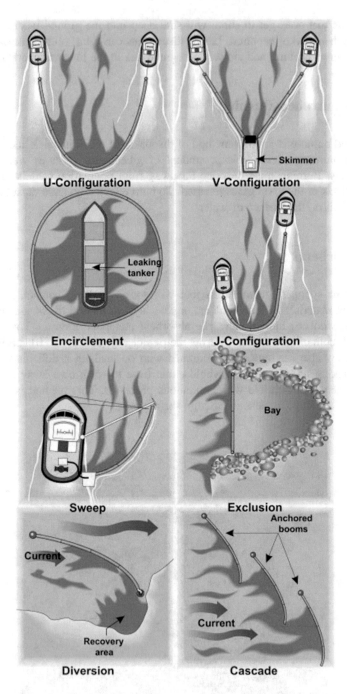

Fig. 9.A.3 Schematics of various configurations used in boom deployment (adapted from Fingas 2011)

9.A.4 Skimmers

Skimmers are floating devices that "skim" oil off the surface. Figure 9.A.3 shows three different forms of oleophilic surface skimmers and their oil collection wells. Their efficiency is dependent upon favorable conditions including calm waters and slow oil speeds. They are prone to clogging with trash in the water and cannot be

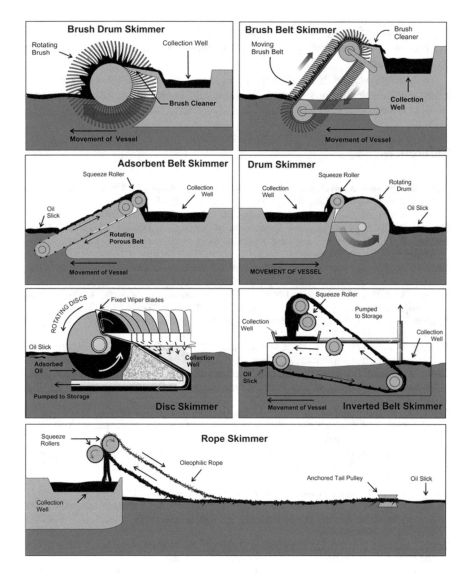

Fig. 9.A.4 Schematics of the principles of three different forms of oleophilic surface skimmers (adapted from Fingas 2011)

used in rough water. In the case of absorbent belt skimmers, the collected oil needs to be physically removed from the surface during oil removal. If the oil is not removed, surfaces could become fouled, making recycling of the surfaces impossible.

9.A.5 Booms

Booms are long floating objects that can be dragged across the water to corral oil and prevent it from spreading. Figure 9.A.4 shows various configurations used in boom deployment.

To sum up, various methods used today, use toxic chemicals which are environmentally unfriendly or slow and tedious requiring significant time and expense.

References

Bayat, A., Aghamiri, S., Moheb, A., and Vakili-Nezhaad, G. R. (2015), "Oil Spill Cleanup from Sea Water by Sorbent Materials," *Chem. Eng. Technol.* **28**, 1525–1528.

Bhushan, B. (2013a), *Introduction to Tribology*, second ed., Wiley, New York.

Bhushan, B. (2013b), *Principles and Applications of Tribology*, second ed., Wiley, New York.

Bhushan, B. (2017), *Nanotribology and Nanomechanics: An Introduction*, fourth ed., Springer International, Cham, Switzerland.

Bhushan, B. (2018), *Biomimetics: Bioinspired Hierarchical-Structured Surfaces for Green Science and Technology*, third ed., Springer International, Cham, Switzerland.

Bhushan, B. (2019a), "Lessons from Nature for Green Science and Technology: An Overview and Bioinspired Superliquiphobic/philic Surfaces," *Phil. Trans. R. Soc. A* **377**, 20180274.

Bhushan, B. (2019b), "Bioinspired Oil-water Separation Approaches for Oil Spill Clean-up and Water Purification," *Phil. Trans. R. Soc. A* **377**, 20190120.

Bhushan, B. and Martin, S. (2018), "Substrate-independent Superliquiphobic Coatings for Water, Oil, and Surfactant Repellency: An Overview," *J. Colloid Interface Sci.* **526**, 90–105.

Brown, P. S. and Bhushan, B. (2015a), "Mechanically Durable, Superoleophobic Coatings Prepared by Layer-by-layer Technique for Anti-smudge and Oil-water Separation," *Sci. Rep.-Nature* **5**, 8701.

Brown, P. S. and Bhushan, B. (2015b), "Bioinspired, Roughness-induced, Water and Oil Super-philic and Super-phobic Coatings Prepared by Adaptable Layer-by-layer Technique," *Sci. Rep.-Nature* **5**, 14030.

Brown, P. S. and Bhushan, B. (2016), "Bioinspired Materials for Water Supply and Management: Water Collection, Water Purification and Separation of Water from Oil," *Phil. Trans. R. Soc. A* **374**, 20160135.

Chandan, R. (1997), *Dairy Based Ingredient*, Eagan Press, St. Paul, MN.

Chauhan, P., Kumar, A., and Bhushan, B. (2019), "Self-cleaning, Stain-resistant and Anti-bacterial Superhydrophobic Cotton Fabric Prepared by Simple Immersion Technique," *J. Colloid Interface Sci.* **535**, 66–74.

Dave, D. and Ghaly, A. (2011), "Remediation Techniques for Marine Oil Spills: A Critical Review and Comparative Analysis," *Am. J. Environ. Sci.* **7**, 423–440.

Etkin, D. S. and Welch, J. (1997) "Oil Spill Intelligence Report International Oil Spill Database: Trends in Oil Spill Volumes and Frequency," *Int. Spill Conf.* 949–952.

Feng, L., Zhang, Z., Mai, Z., Ma, Y., Liu, B., Jiang, L., and Zhu, D. (2004), "A Super-Hydrophobic and Super-Oleophilic Coating Mesh Film for the Separation of Oil and Water," *Angew. Chem., Int. Ed.* **43**, 2012–2014.

Fingas, M. (ed.), (2011), *Oil Spill Science and Technology: Prevention, Response, and Cleanup*, Elsevier, Amsterdam.

Gao, S. J., Shi, Z., Zhang, W. B., Zhang, F., and Jin, J. (2014), "Photoinduced Superwetting Single-walled Carbon Nanotube/TiO2 Ultrathin Network Films for Ultrafast Separation of Oil-in-Water emulsions," *ACS Nano* **8**, 6344–6352.

Harkins, W. D. and Brown, F. E. (1919), "The Determination of the Surface Tension (Free Surface Energy), and the Weight of Falling Drops: the Surface Tension of Water and Benzene by the Capillary Height Method, *J. Am. Chem. Soc.* **41**, 499–524.

Hsieh, C., Change, B., and Lin, J. (2011), "Improvement of Water Oil Repellency on Wood Substrate by Using Fluorinated Silica Nano Coating," *Appl. Surf. Sci.* **257**, 7997–8002.

ITOPF (2015), "Oil Tanker Spill Statistics," see http://www.itopf.com/knowledge-resources/data-statistics/statistics.

Kokal, S. L. (2005), "Crude Oil Emulsions: A State-of-the-Art Review," *SPE Production & Facilities* **20**, 5–13.

Kondash, A. and Vengosh, A. (2015), "Water Footprint of Hydraulic Fracturing," *Environ. Sci. Technol. Lett.* **2**, 276–280.

Li, F., Bhushan, B., Pan, Y., and Zhao, X. (2019a), "Bioinspired Superoleophobic/ superhydrophilic Functionalized Cotton for Efficient Separation of Immiscible Oil-Water Mixtures and Oil-Water Emulsions," *J. Colloid Interface Sci.* **548,** 123–130.

Li, F., Kong, W., Bhushan, B., Zhao, X., and Pan, Y. (2019b), "Ultraviolet–driven Switchable Superliquiphobic/superliquiphilic Coating for Separation of Oil-water Mixtures and Emulsions and Water Purification," *J. Colloid Interface Sci.* **557**, 395–407.

Liu, K., Cao, M., Fujishima, A., and Jiang, L. (2014), "Bio-inspired titanium dioxide materials with special wettability and their applications," *Chem. Rev.* **114**, 10044–10094.

Liu, L., Pan, Y., Bhushan, B., and Zhao, Z. (2019), "Mechanochemical Robust, Magnetic-Driven, Superhydrophobic 3D Porous Materials for Contaminated Oil Recovery," *J. Colloid Interface Sci.* **538**, 25–33.

Martin, S., Brown, P. S., and Bhushan, B. (2017), "Fabrication Techniques for Bioinspired, Mechanically-durable, Superliquiphobic Surfaces for Water, Oil, and Surfactant Repellency," *Adv. Colloid Interf. Sci.* **241**, 1–23.

Nanda, D., Sahoo, A., Kumar, A., Bhushan, B. (2019), "Facile Approach to Develop Durable and Reusable Superhydrophobic/superoleophilic Coatings for Steel Mesh Surfaces," *J. Colloid Interface Sci.* **535**, 50–57.

Nwaizuzu, C., Joel, O. F., and Sikoki, F. D. (2015), "Evaluation of Oil Spill Dispersants with a focus on their toxicity and Biodegradability," Society of Petroleum. Engineers (https://doi.org/10.2118/178321-ms).

Ohama, Y. and Van Gemert, D. (Eds.). (2011), *Application of Titanium Dioxide Photocatalysis to Construction Materials: State-of-the-Art Report of the RILEM Technical Committee 194-TDP* (Vol. 5). Springer Science and Business Media, Heverlee, Belgium.

Panda, A., Varshney, P., Mohapatra, S. S., Kumar, A. (2018), "Development of Liquid Repellent Coating on Cotton Fabric by Simple Binary Silanization with Excellent Self-cleaning and Oil-water Separation Properties," *Carbohydr. Polym.* **181**, 1052–1060.

Prince, L. (2012), *Microemulsions: Theory and Practice*, Elsevier, New York.

Przybylak, M., Maciejewski, H., Dutkiewicz, A., Dabek, I., Nowicki, M. (2016), "Fabrication of Superhydrophobic Cotton Fabrics by a Simple Chemical Modification," *Cellulose* **23**, 2185–2197.

Ramseur, J. L. (2015), "Deepwater Horizon Oil Spill: Recent Activities and Ongoing Developments," see https://www.fas.org/sgp/crs/misc/R42942.pdf.

Rao, D. G., Senthilkumar, R., Byrne, J. A., and Feroz, S. (2012), *Wastewater Treatment: Advanced Processes and Technologies*, CRC Press, Boca Raton, Florida.

Rumble, J. R. (2018), *CRC Handbook of Chemistry and Physics*, 99th ed., CRC Press, Boca Raton, FL.

Speight, J. (2017), *Environmental Organic Chemistry for Engineers*, Butterworth-Heinemann, Oxford, UK.

Takamura K., Fisher, H., and Morrow, N. R. (2012), "Physical Properties of Aqueous Glycerol Solutions", *J. Pept. Sci. Eng.* **98–99**, 50–60.

Tudu, B. K., Kumar, A., and Bhushan, B. (2019), "Fabrication of Superoleophobic Cotton Fabric for Multi-purpose Applications," *Phil. Trans. R Soc. A* **377**, 20190129 (2019).

Wang, H., Zhou, H., Liu, S., Shao, H., Fu, S., Rutledge, G. C., and Lin, T. (2017), "Durable, Self-healing, Superhydrophobic Fabrics from Fluorine-free, Waterborne, Polydopamine/alkyl Silane Coatings", *RSC Adv.* **7**, 33986–33993.

Xue, Z., Wang, S., Lin, L., Chen, L., Liu, M., Feng, L., and Jiang, L. (2011), "A Novel Superhydrophilic and Underwater Superoleophobic Hydrogel-Coated Mesh for Oil/Water Separation," *Adv. Mater.* **23**, 4270–4273.

Zeng, X., Qian, L., Yuan, X., Zhou, C., Li, Z., Cheng, J., Xu, S., Wang, S., Pi, P., and Wen, X. (2016), "Inspired by Stenocara Beetles: from Water Collection to High-Efficiency Water-in-Oil Emulsion Separation," *ACS Nano* **11**, 760–769.

Zhu, G., Pan, Q., and Liu, F. (2011), "Facile Removal and Collection of Oils from Water Surfaces through Superhydrophobic and Superoleophilic Sponges," *J. Phys. Chem. C* **115**, 17464–17470.

Zhu, T., Li, S., Huang, J., Mihailiasa, M. and Lai, Y. (2017), "Rational Design of Multi-layered Superhydrophobic Coating on Cotton Fabrics for UV Shielding, Self-cleaning and Oil-water Separation," *Mater. Des.* **134**, 342–351.

Chapter 10
Closure

Access to a safe water supply is a basic human right. Fresh water sustains human life and is vital for human health. Water availability depends upon the amount of water physically available, and whether it is safe for human consumption. Some of the arid regions of the world lack adequate safe drinking water. It is estimated that about 800 million people worldwide lack basic access to drinking water (WHO/UNICEF 2017; Bhushan 2020). This means that they cannot reach a protected source of drinking water within a total walking distance of 30 min. About 2.2 billion people (nearly a third of the global population) do not have access to a safe water supply, meaning no drinking water on the property that is available at all times and free of contamination. Almost half of people drinking from water from unprotected sources live in sub-Saharan Africa. Women and girls regularly experience discrimination an inequalities in the rights of safe drinking water.

Over 2 billion people, representing nearly one third of the world live in countries experiencing high water stress, meaning that water resources consumed are not regenerated to the necessary extent by rain or the return of the purified water (UN 2018; Bhushan 2020). About 4 billion people experience water scarcity during at least one month of the year (WWAP 2019; Bhushan 2020). Due to climate change, bad economics, and/or poor infrastructure in some parts of the world, the number of people experiencing high water stress continues to increase. The majority of the people affected by water stress live in North Africa, the Middle East, and South Asia but the people living in North and South America, including the Southwest United States are also increasingly affected.

Water requirements to meet basic water needs is about 7.5–15 L day^{-1}, typically 50 L day^{-1} (Bhushan 2020). Water consumption continues to grow worldwide driven by a combination of population growth, socio-economic development, and rising demand in the industrial and domestic sectors. Current supply of fresh water needs to be supplemented to meet future needs.

Roughly 70% of the Earth's surface is covered by water, however, only 0.79% (11 quadrillion m^3) of all water is found as surface water in lakes and rivers (0.03%) and groundwater (0.76%) (Bhushan 2018, 2019, 2020). The distribution of this water

© Springer Nature Switzerland AG 2020
B. Bhushan, *Bioinspired Water Harvesting, Purification, and Oil-Water Separation*, Springer Series in Materials Science 299, https://doi.org/10.1007/978-3-030-42132-8_10

is not uniform across the world, for example, with around 20% found in the North American Great Lakes. In addition to all water on earth, the earth's atmosphere contains about 0.001% of global water (13 trillion m^3 or 1.3×10^{16} L). It is found in the form of clouds, fog, mist, and water vapor. It is a small amount in global terms, but it is about 3.3% of water in rivers and lakes. It could be a large source to supplement safe and drinking water supply.

Living nature provides many lessons for harvesting water from the atmosphere. Plants and animals have evolved species, which can survive in the most arid regions of the world by water collection from fog and condensation in the night. Before the collected water evaporates, species have developed mechanisms to transport water for storage or consumption. These species possess unique chemistry and structures on or within the body for collection and transport of water. By studying the lessons from nature (both plants and animals), for example, beetles, cactus, and spider silk, grass, various surface structures and chemistries can be developed for efficient water harvesting from fog and condensation.

Bioinspired water harvester can be used for water harvesting from fog and/or condensation of water vapor for the use of a household or a small community. In addition, these harvesters can be used in various emergency and defense applications (Bhushan 2019, 2020). Emergency applications, such as natural disasters, could benefit for short periods from portable units which could be dropped from air. The cost of clean water in military bases in a combat zone can be high and water harvesters become attractive. Various designs for large and portable 3D towers have been developed.

Bioinspired water desalination and water purification approaches are also discussed. Desalination of water from the oceans or brackish groundwater is essential for some parts of the world for water supply. With population growth and industrial revolution, contamination of water supply is a concern. Living nature provides examples of membranes for liquid purification. It is hoped that by taking inspiration from the chemistries and surface structures of these natural species, bioinspired methods can be created that can improve desalination as well as water purification from contamination.

Water contamination by human activity and unsafe industrial practices, as well as population growth continues to increase. Water contamination is one of the major environmental and natural resource concerns in the twenty-first century, including North America. Oil contamination can occur during operation of machinery, oil exploration and transportation, and due to operating environment. Oil spills occasionally occur during oil exploration and transportation. Water contamination with various chemicals is also a major concern with growing population and unsafe industrial practices of waste disposal. Commonly used oil-water separation techniques for oil-spill cleanup are either time consuming, energy intensive, and/or environmentally unfriendly. Bioinspired superhydrophobic/superoleophobic and superoleophobic/superhydrophilic surfaces have been developed which are sustainable and green or environmentally friendly. Bioinspired oil-water separation techniques can be used to remove oil contaminants from both immiscible oil-water mixtures and oil-water emulsions. Coated porous surfaces with an affinity to water and repellency to oil and vice versa are commonly used. The former combination of affinity to water and repellency to oil is preferred to avoid oil contamination of the porous substrate. Oil-water emulsions require porous materials with a fine pore size. Recommended porous materials include steel mesh and cotton fabric for immiscible oil-water mixtures and cotton for oil-water emulsions.

Bioinspired approaches are attractive to develop materials and surfaces in an environmentally friendly and sustainable manner to supplement water supply and remove contamination.

References

Bhushan, B. (2018), *Biomimetics: Bioinspired Hierarchical-Structured Surfaces for Green Science and Technology*, third ed., Springer International, Cham, Switzerland.

Bhushan, B. (2019), "Bioinspired Water Collection Methods to Supplement Water Supply," *Phiosl. Trans. R. Soc. A* **377**, 20190119.

Bhushan, B. (2020), "Design of Water Harvesting Towers and Projections for Water Collection from Fog and Condensation," *Phiosl. Trans. R. Soc. A* **378**, 20190440.

UN (2018), *Sustainable Development Goal 6: Synthesis Report 2018 on Water Sanitation*, United Nations, New York. www.unwater.org/app/uploads/2018/07/SDG6_SR2018_web_v5.pdf.

WHO/UNICEF (2017), *Progress on Drinking Water, Sanitation and Hygiene: 2017 Update and SDG Baselines*, World Health Organization (WHO) and the United Nations Children's Fund (UNICEF), Geneva, Switzerland. See https://washdata.org/sites/default/files/documents/reports/2018-01/JMP-2017-report-final.pdf.

WWAP (2019), *The United Nations World Water Development Report 2019: Leaving No One Behind*, WWAP (UNESCO World Water Assessment Programme), UNESCO, Paris, France.

Index

A

A decanoic acid (DA), 170, 171
Adhesive properties, 54
Adsorbents, 161, 162
Amino acids, 166, 167
Amphiphilic, 163, 166
Analytical balance, 72–74
Anisotropic wetting, 55
Aquaporins, 163, 165–169, 171
Arid, 11–14, 16–18, 20, 22, 25, 33, 35
Array, 64, 68, 69, 71, 72, 93, 94, 100, 106, 107, 111
Atacama desert, 14, 25
Atomic Force Microscope (AFM), 186, 188
Awns, 51

B

Ball-on-flat tribometer, 187, 188
Barbs, 49, 51, 55
Beads on a string, 54
Beetle, 20, 34–36, 40
Beetle-inspired surfaces, 64, 65, 75–78, 81, 98, 99
Big Sur, CA, 17, 18
Bioinspired, 5, 9, 10
Bioinspired surfaces, 156, 158
British Petroleum, 7
B. subtilis, 164

C

Cactus, 48, 51, 55, 56
California central coast, 17, 18, 20
Capillary force, 52, 53
Carbon Nanotubes (CNT), 163, 168, 169, 171, 172

Carlsbad Desalination Plant, 5
Cassie-Baxter state, 175
Catalysis, 162, 167
Centerline, 72, 82–84, 96
Chirality, 166
Coalescence, 122, 126, 127, 130, 135, 138, 144, 145, 147, 148, 150, 151
Coastal deserts, 11, 13, 16, 17, 25, 36
Cold deserts, 11, 13, 26
Colloid, 181
Condensation, 3, 8, 9, 14, 16, 21, 22, 34, 63, 64, 69, 74, 75, 100, 104, 106, 108, 109
Conical array, 64, 68, 70, 75, 81, 85, 93, 99, 110, 111
Conical geometry, 63, 66, 67, 75, 78
Conical surfaces, 63, 64, 66, 67, 75, 78, 79, 87, 94, 110
Contact angle, 175–177, 183, 184, 187, 193–198, 200, 201, 203–205, 210
Contact angle hysteresis, 175
Contaminants, 161, 162, 164, 168, 171
Contamination, 1, 7–10, 225–227
Copiapoa haseltoniana, 48
Crane, EE-3186, 119, 120
Creosote bushes, 25, 26, 28
Creosote Larrea tridentata, 52
Curvature gradient, 82–84, 90, 94–97
Cylinder, 66, 75, 78–80, 100–102
Cylindrical P4VP, 169

D

Deepwater Horizon oil spill, 7
Defense applications, 155, 158
Deionized water, 201–203
Desalination, 5, 9, 10, 161, 162, 164, 171, 172

© Springer Nature Switzerland AG 2020
B. Bhushan, *Bioinspired Water Harvesting, Purification, and Oil-Water Separation*, Springer Series in Materials Science 299, https://doi.org/10.1007/978-3-030-42132-8

Printed in the United States
by Baker & Taylor Publisher Services